SECOND CATALOGUE OF X-RAY SOURCES

SECOND CATALOGUE

OF

X-RAY SOURCES

by

P. R. AMNUEL, O. H. GUSEINOV, and YU. RAKHAMIMOV

Reprinted from

Astrophysics and Space Science, Vol. 82, No. 1

D. Reidel Publishing Company

P.O. Box 17, 3300 AA Dordrecht, Holland

ISBN-13: 978-90-277-9055-2   e-ISBN-13: 978-94-009-7964-2
DOI: 10.1007/ 978-94-009-7964-2

Order Ref. No. 90−277−9055−2

SECOND CATALOGUE OF X-RAY SOURCES*

P. R. AMNUEL, O. H. GUSEINOV, and SH. YU. RAKHAMIMOV

Shemakha Astrophysical Observatory, Azerb. SSR, U.S.S.R.

(Received 21 August, 1980; in revised form 23 June, 1981)

Abstract. A catalogue of X-ray sources containing 677 objects known as of 1979 September has been compiled. X-ray data collected by means of rockets, balloons, and satellites during 1964–1979 are listed. Optical and radio counterparts are suggested. The existence of the weak X-ray sources population in the Galaxy is indicated. The number of X-ray sources of each type in the Galaxy is estimated.

1. Introduction

Recently, we compiled a catalog of X-ray sources containing 517 objects (Ammuel *et al.*, 1979). After the preparation of the catalog surveys by OSO-7, Ariel-5 and partially by HEAO-1 satellites were finished, in general, these surveys confirmed Uhuru results (Forman *et al.*, 1978) and added ~100 weak X-ray sources in the 1–20 keV range. Nearly 40 sources in hard (more than 20 keV) and soft (less than 1 keV) ranges were discovered. The second version of our catalog (2 XRS catalog) contains 677 objects without taking into account recent investigations by HEAO-2. The sky survey is practically complete to X-ray intensities of nearly 4 μJy in the 2–6 keV range. The data on 677 sources are the results of X-ray sky surveys by rockets, balloons and satellites of the first generation. Now the second-generation satellite HEAO-2 has been launched. The sets of HEAO-2 have a sensitivity to $\sim0.03\,\mu$Jy when the exposure time is approximately 10^3 s. The first HEAO-2 results show that newly discovered sources have soft spectra. These sources are connected with radiation of the stellar coronae and other objects of a nonrelativistic nature, as opposed to earlier-discovered sources which possibly are to be connected, in general, with processes of radiation in the neighborhood of condensed stars (neutron stars, black holes and partly white dwarfs). Then the interpretation of the first-generation satellites data is important.

The catalog (Table I) is divided into 10 columns. Column (1) gives the source name. Each listing is truncated to the first decimal place in minutes of right ascension and to the first decimal place in degrees of declination. An 2 XRS designates a source from this catalog.

Column (2) lists other names of the sources.

Column (3) contains the position of the maximum probability density in right ascension and declination (in 1950 coordinates). From all the measurements we selected the coordinates whose error boxes have the least area.

Column (4) contains the position of the maximum probability density in galactic coordinates (l^{II}, b^{II}).

---

* Astrophysics and Space Science review paper.

Astrophysics and Space Science **82** (1982) 3–103. 0004–640X/82/0821–0003$15.15.

Column (5) gives the area of the error box in square degrees.

Column (6) lists the dates of the measurements. The data for the majority of the measurements of each source are indicated. If a source was not investigated extensively by means of satellites, then the majority of the rocket flight data are recorded. For well-investigated sources, such as Cyg X-1, Her X-1 and Sco X-1, only the general satellite data are noted.

Column (7) gives the type of measuring system used (R = rocket and B = balloon; for the satellites U = Uhuru, A = Ariel-5, S3 = SAS-3, O7 = OSO-7, O8 = OSO-8, H = HEAO-1, V5 = Vela-5, C = Copernicus, Cos = Cosmos 428, Sal = Salut 4, and Ari = Ariabata).

Column (8) gives the range (in keV) at which the measurement was performed.

Column (9) gives the maximum intensity for a variable X-ray source or the average intensity for a constant source. The intensities are converted to the 2–6 keV range and are given in μJy. For the conversion, we assumed that a source has a Crab-like spectrum. The source spectrum and the Crab-like spectrum usually differed by no more than a factor of 1.5–2 in intensity. For the Crab-like spectrum 1 Uhuru counts s^{-1} correspond to 1.7×10^{-11} ergs $cm^{-2} s^{-1}$, 1 μJy corresponds to 9.6×10^{-12} ergs $cm^{-2} s^{-1}$ or to 0.6 Uhuru counts s^{-1} (in the 2–6 keV range). The intensities of X-ray sources observed in the soft (less than 1 keV) and hard (more than 20 keV) ranges are listed in Table II. Data for some X-ray sources observed by rockets were actually confused.

Column (10) gives the ratio of the maximum observed value of intensity to the minimum one for variable sources. Column (10) also gives additional comments on the nature of the source (bursts, transients, soft and hard sources, etc.).

Table II gives more detailed information, such as data for X-ray spectra, variability, and possible identification. Table II also gives references for each X-ray source.

Table III gives general data on X-ray sources identified with binary systems and divided into 2 groups: massive and low-mass systems.

Table IV contains data on 17 X-ray pulsars (radiating in the 1–20 keV range).

Table V lists 159 X-ray sources associated with extragalactic objects (clusters of galaxies, quasars, emission galaxies, etc.).

Table VI lists 40 transient X-ray sources.

Table VII lists 45 X-ray sources whose observed bursts last less than 100 s. These sources are designated as SB (bursters of short duration).

Table VIII lists 18 bursters of long duration (LB) for which the burst duration is no more than several hours.

Table IX lists 24 X-ray Supernova Remnants.

Table X lists 44 soft X-ray sources observed in the range of less than 1 keV (without taking into account Supernova Remnants).

Table XI lists 8 hard sources observed in the range of more than 20 keV.

So, the 2 XRS catalog contains additionally to 1 XRS data on 116 weak X-ray sources (range 1–20 keV, intensities less than 20 μJy) discovered during last surveys, 31 soft and 2 hard sources and 11 X-ray Supernova Remnants.

2. Bright X-ray Sources

There are not sufficient new data on bright X-ray sources (intensities of more than $20\,\mu$Jy). Matilsky (1977), Meurs (1978), Rothenflug *et al.* (1979), Protheroe and Wolfendale (1980) have investigated the distribution of bright X-ray sources in the Galaxy on the basis of the Uhuru data. The problem was also suggested in works by Amnuel *et al.* (1979) and Amnuel and Guseinov (1980) based on data of the 1 XRS catalog. The main conclusions are summarized below:

Bright sources divide into two types according to their spectral hardness and identifications. Hard, steady and transient sources ($kT \gtrsim 15\,\text{keV}$) associate with massive binary systems (Cyg X-1, Cen X-3, etc.). For these sources a horizon is located at a distance of $\sim 7\,\text{kpc}$, the observed number is 27, and the expected number in the Galaxy is ~ 100–200. The sources are located in the ring around the galactic center as other massive objects and the gas component distribute. The average X-ray luminosity (2–6 keV range) is $\sim 10^{37}\,\text{erg s}^{-1}$. A component of a massive star is a young neutron star with a strong magnetic field. In rarer cases, a component may be a black hole.

The soft sources (steady, transient and bursters) with $kT \lesssim 5\,\text{keV}$ are, in general, connected with low-mass binary systems (Sco X-1, Cyg X-2, etc.). A horizon for soft sources is located at $\sim 10\,\text{kpc}$, the observed number is 40 (steady sources), and the expected number in the Galaxy is ~ 100–200. The distribution of soft sources in the Galaxy will be written as $n \simeq n_0 \times 10^{-(0.12-0.16)R}$, where n_0 and n are the surface density of source numbers in the galactic center region and at the distance R kpc from the galactic center. The average X-ray luminosity is $\sim 5 \times 10^{37}\,\text{erg s}^{-1}$ in the 2–6 keV range. The normal component in these systems is a dwarf star with a mass of less than $2M_\odot$, and the X-ray component is a condensed star (neutron star, in rare cases, a black hole). According to Amnuel and Guseinov (1980), the low mass X-ray systems are binaries with a massive primary star and an initial extremely small mass ratio $q \lesssim 0.1$. The time of such systems, evolution before Roche overflow is more than 10^9 yr.

The average parameters of other types of X-ray sources (transient, bursters, etc.) are suggested in 1 XRS catalog and are not sufficiently different. So we will suggest below a problem of the nature of weak X-ray sources (intensities less than $20\,\mu$Jy) nonidentified with extragalactic objects.

3. Weak X-ray Sources

The 2 XRS catalog contains 275 sources with intensities of less than $20\,\mu$Jy nonidentified with extragalactic objects. Fifteen sources of that type associate with galactic objects: early type stars (2 XRS $00536 + 604$, $01147 + 650$, $07574 - 484$, $10440 - 594$, $10528 + 606$), flare stars of U Gem type (2 XRS $06270 + 216$, $07284 + 060$, $09483 + 121$, $22116 + 124$), Novae and novalike stars (2 XRS $03277 + 432$, $21407 + 433$), binary semi-detached system β Per, binary system with magnetic white dwarf AM Her, peculiar object SS 433 (2 XRS $19094 + 047$) and globular cluster M15 (2 XRS $21275 + 119$).

The problem of the physical nature of weak X-ray sources was suggested by many

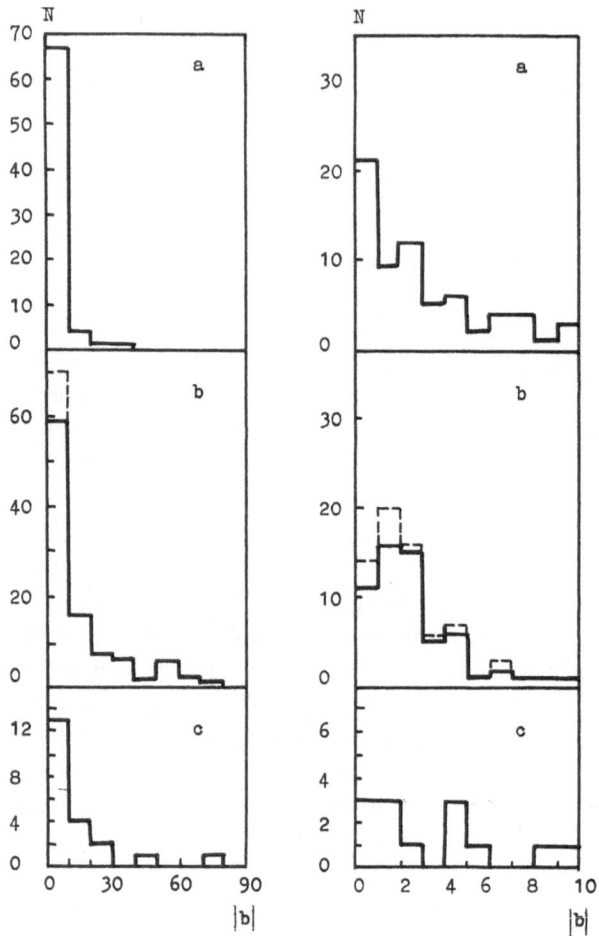

Fig. 1.   (a) Distribution of bright X-ray sources ($S > 20\,\mu$Jy) on their galactic latitudes; (b) Distribution of weak X-ray sources ($S = 4$–$20\,\mu$Jy) on their galactic latitudes. The expected numbers of sources taking into account confusion effects are marked by dotted line; (c) Distribution of nova stars at distances to 1.5 kpc on their galactic latitudes. Distributions with intervals of 10° are shown on the left, distribution at $|b| < 10°$ with intervals of 1° åre shown on the right.

authors. Gursky and Schreier (1975) suggested that all these sources are extragalactic. Matilsky *et al.* (1972), Holt *et al.* (1974), and Murray (1977) did not specify the galactic or extragalactic nature of these sources. Now the presence of weak X-ray sources in the Galaxy discussed in works by Amnuel and Guseinov (1974) and Amnuel *et al.* (1979) is proved, but their number, distribution and physical nature is a problem.

    The latitude distribution of weak X-ray sources nonidentified with extragalactic objects (WS) is shown in Figures 1b and 2b. WS are divided into two groups: WS with intensities of 4–20 μJy and with intensities of less than 4 μJy. In the first case the 2 XRS catalog is practically complete. The excess of sources at low latitudes $|b| < 30°$ is seen. The value of excess is approximately 90–100 sources.

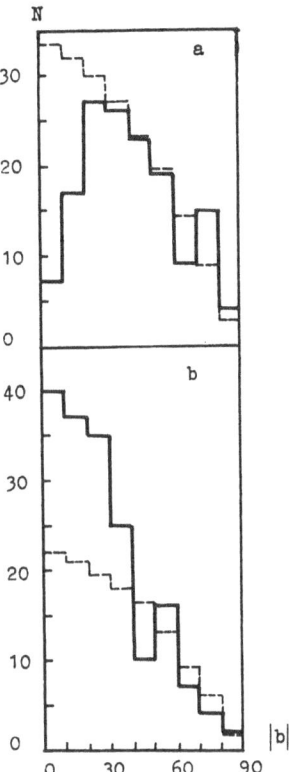

Fig. 2. (a) Distribution of X-ray sources associated with extragalactic objects on their galactic latitudes. The expected uniform distribution in suggestion, that sources observed higher than 30° are located uniformly, is shown by a dotted line; (b) the analogous distribution of weak sources ($S < 4\,\mu$Jy).

If one does not take into account the presence of the population of weak sources in our Galaxy, then two possibilities are to be suggested:

(1) The sources are extragalactic objects, not associated because of their proximity to the galactic plane;

(2) the sources are distant bright sources (L_x more than $10^{36}\,\mathrm{erg\,s^{-1}}$).

For the exclusion of the first possibility, let us suggest that we observe the number of extragalactic sources and WS together with their expected number, suggesting that the distribution of sources at latitudes $|b| > 30°$ is to be uniform, see Figure 2. The deficit of extragalactic sources at low latitudes is seen. This deficit is to be decreased at the expense of WS. This decreases the value of excess WS at low latitudes. However, this does not totally destroy this excess.

The presence of distant bright sources also does not explain this excess. Even if $|\bar{Z}| \sim 0.4\,\mathrm{kpc}$ for bright sources (Amnuel and Guseinov, 1980) and distance from the Sun is more than 12 kpc, then one expects the excess only at $|b| < 2°$. The investigations of WS distribution at $|b| < 10°$ do not show sufficient excess at $|b| < 2°$, even if one takes into account confusion by bright sources.

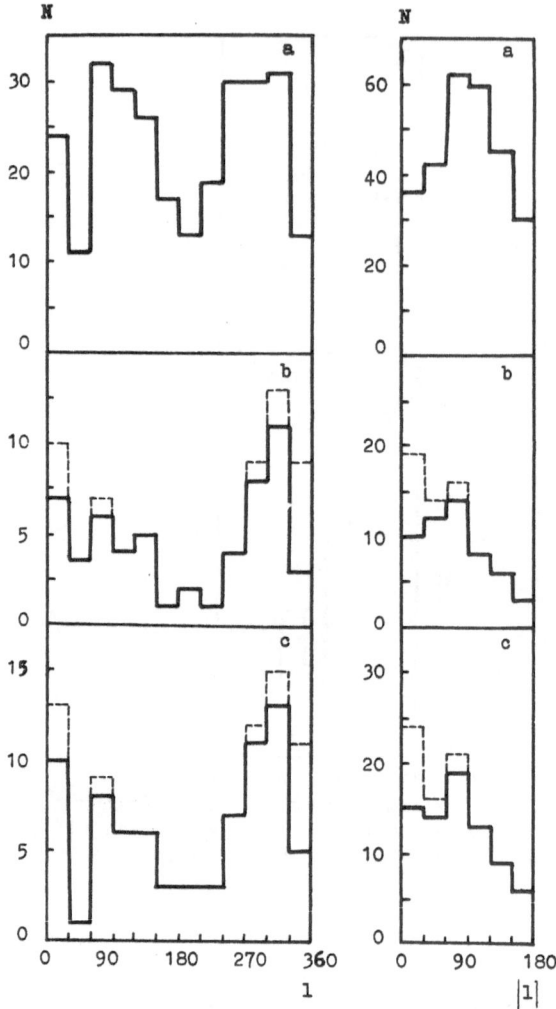

Fig. 3. Distribution of sources on their galactic longitudes (left) and symmetrized distribution on angular distances from the direction towards to the galactic center (right). (a) all weak sources unassociated with extragalactic objects (WS); (b) WS with $S \geqslant 4\,\mu$Jy located at $|b| < 5°$; (c) WS with $S \geqslant 4\,\mu$Jy located at $|b| < 20°$. Expected distributions with taking into account confusion effects are marked by a dotted line.

A special interest is in the investigation of the longitude distribution of WS. The sufficient excess of WS at an angular distance 60–120° from the direction towards the galactic center is seen in Figure 4a. The value of sources in excess is ∼100 (as compared with sources number at $|l| = 120$–180°).

For further investigations we divided all WS into two groups: (i) Intensities $S \geqslant 4\,\mu$Jy and (ii) $S < 4\,\mu$Jy. In case (i) the 2 XRS catalog is practically complete. Figures 1b, 4b, and 4c show longitude and latitude distributions of WS with $S \geqslant 4\,\mu$Jy. Expected distributions taking into account confusion effects are also shown. A strong excess of

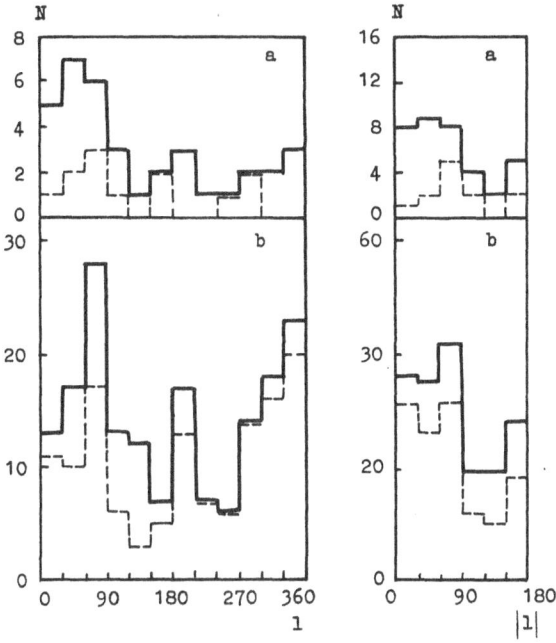

Fig. 4. Distribution of novae (a) and stars of U Gem-type (b) on their galactic longitude (left) and on their angular distance from the direction towards the galactic center (right). (a) novae located at distances less than 2 kpc are marked by solid line, that at distances less than 1 kpc – by dotted line; (b) stars of U Gem-type located at distance less than 0.5 kpc are marked by solid line, that at distances less than 0.3 kpc – by dotted line.

such WS at $|b| < 10°$ is seen. As to the longitude excess, if one takes into account the confusion effects, then the expected numbers of WS ($S \geqslant 4\,\mu$Jy) at $|b| < 5°$ and $|b| < 20°$ show an increase in the galactic center direction.

Bright sources concentrate towards the galactic center direction much more than WS (taking into account the confusion effects). One does not totally exclude the presence of bright distant sources among the WS ($S \geqslant 4\,\mu$Jy), but the fraction of these sources is not great.

The ratio N_c/N_{ac} (N_c and N_{ac} are numbers or sources at longitudes from 270 to 90° and from 90 to 270° accordingly) for WS ($S \geqslant 4\,\mu$Jy) is near 1.7. If one takes into account the confusion effects, then N_c/N_{ac} is 2.1. For the bright sources this ratio is 19. So, WS ($S \geqslant 4\,\mu$Jy) are really a population of close and weak X-ray sources. Thus two hypotheses are to be discussed:

(i) Some weak X-ray sources are associated with novae and nova-like stars (GK Per, SS Cyg), and other systems with white dwarfs. One suggests a possible connection of WS ($S \geqslant 4\,\mu$Jy) with such systems.

(ii) WS ($S \geqslant 4\,\mu$Jy) are connected with systems of low mass normal stars and condensed ones – such systems may arise during the stage preceding the bright X-ray source stage (Amnuel *et al.*, 1974).

Let us consider assumption (i). The longitude and latitude distributions of novae located at distances of 2 kpc from the Sun are shown in Figures 1c and 4a. Distributions are similar to the observed ones for WS ($S \geqslant 4\,\mu$Jy). Amnuel and Guseinov (1980) showed that novae distribute in the Galaxy as $n \simeq n_0 \times 10^{-mR}$. For novae located in the solar neighborhood, Amnuel and Guseinov show that $m = 0.22$–0.24, while at $R \lesssim 6$–7 kpc the value of m increases to $m \approx 0.5$. If $m = 0.23$, one finds that N_c/N_{ac} = 1.7–2.1 (as for WS with $S \geqslant 4\,\mu$Jy) corresponds to the horizon at the distance of 1.2 kpc. Then $|\bar{Z}| \approx 0.2$ kpc which corresponds to observed values of $|\bar{Z}|$ for novae. The value of WS ($S \geqslant 4\,\mu$Jy) in excess at low latitudes is ≈ 30–40 objects. Then n ($R = 10$ kpc) ≈ 6–8 kpc^{-2}, $\bar{L}_x \approx 10^{34}$ erg s^{-1}. This value of $\bar{L}_x$ is more than that for WS associated with novae and related stars (for SS Cyg $\bar{L}_x \approx 2 \times 10^{32}$ erg s^{-1}). If WS ($S \geqslant 4\,\mu$Jy) distribute as novae, then the total expected number of such sources in the Galaxy is $\sim 10^5$. The value of the X-ray background luminosity is $\gtrsim 10^{39}$ erg s^{-1} which is 20 times greater than the observed value (Schwartz, 1979). Besides that, in this case more than half of the X-ray background luminosity arises in the region nearer than 30° to the galactic center direction, which also contradicts the observations. Apparently, a little fraction of WS ($S \geqslant 4\,\mu$Jy) concerns novae and related stars populations.

Then it is worthwhile discussing a hypothesis (Amnuel et al., 1974) which suggests the presence of a weak galactic X-ray sources population connected with condensed stars in low-mass binaries. For such a population $|\bar{Z}| \approx 0.3$–0.4 kpc as for bright X-ray sources. As the excess of WS ($S \geqslant 4\,\mu$Jy) is observed mainly at $|b| < 10°$, then a horizon is located at the distance $\gtrsim 2$ kpc.

One estimates the distance of the horizon according to the value of N_c/N_{ac} if the value of the concentration degree $m = 0.14$–0.16 is correct (Amnuel and Guseinov, 1980). In the last case, the horizon is located at 2 kpc. Then n ($R = 10$ kpc) $= 1$–3 kpc^{-2}, the expected total number of such sources in the Galaxy is $(2$–5$)\,10^3$ and $\bar{L}_x \approx 2 \times 10^{34}$ erg s^{-1}. These estimates do not contradict the galactic X-ray background data. There are 100–200 bright X-ray sources in the Galaxy ($L_x \gtrsim 10^{36}$ erg s^{-1}) with a lifetime of $\sim 5 \times 10^7$ yr. Then the lifetime of such sources in the weak source stage is $(0.5$–2$) \times 10^9$ yr.

Why do WS ($S \geqslant 4\,\mu$Jy) not associate with normal low mass stars? One notes, that such stars are not selected among the field stars. The effects of X-ray emission reflection are not sufficient in this case because of the small X-ray luminosity: the optical luminosity of normal star is $\sim 2 \times 10^{33}$ erg s^{-1}, and the reflected X-ray luminosity is less than 10^{32} erg s^{-1}. The association facilitates if transient phenomenon is observed (see references for 2 XRS 06203 $-$ 002, etc.). Between the flares such stars have $V \gtrsim 19$–20.

The nearest WS located at a distance of less than 1 kpc (~ 10 sources) are to be associated with stars of $V \approx 14$–16, but one could not select these sources from 275.

Let us suggest now distributions of WS ($S < 4\,\mu$Jy) represented in Figures 2b, 5b–d. An excess of such WS at $|b| < 30°$ exists. The sources do not concentrate strongly towards to the galactic plane, but the excess at low latitude is probably real, especially if one takes into account the effects of confusion at $|b| < 5°$. The value of the low-latitude excess is ≈ 40–50 sources.

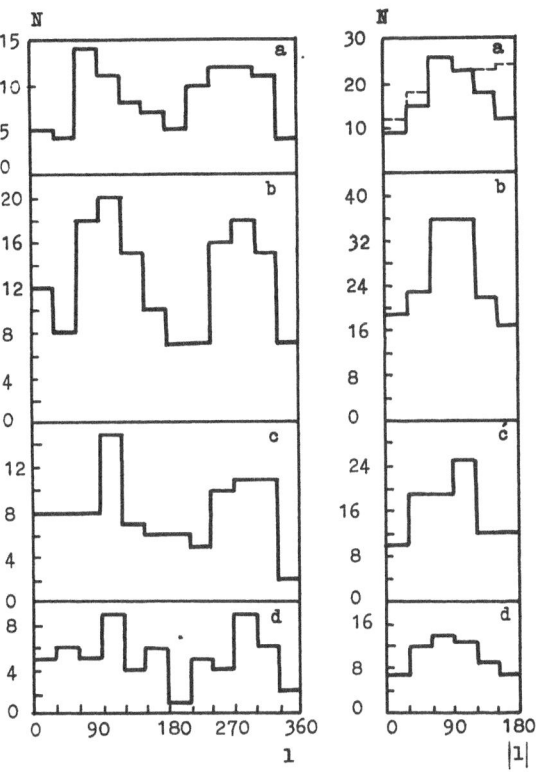

Fig. 5.   Distribution of WS ($S < 4\,\mu$Jy) in their galactic longitude (left) and in their angular distance from the direction towards the galactic center (right). (a) Uhuru sources. Expected distribution taking into account the incompleteness of the survey is shown by solid line; (b) 2 XRS sources at $|b| > 5°$; (c) 2 XRS sources at $|b| > 20°$; (d) 2 XRS sources at $|b| > 30°$.

One would explain such excess by taking into account a great deficit of low latitude identified extragalactic sources (Figure 2a). This deficit may be disrupted at the expense of WS ($S < 4\,\mu$Jy), then the distribution of WS ($S < 4\,\mu$Jy) is to be practically uniform.

The longitude distribution shows an excess at $|l| \approx 60$–$120°$, which is not disrupted by effects of confusion only, as it is for WS ($S \geqslant 4\,\mu$Jy).

It is natural to conclude that WS ($S < 4\,\mu$Jy) are to be, in the main, nonidentified extragalactic objects. The presence of longitude excess can be explained by several reasons:

(i) The incompleteness of Uhuru survey for WS ($S < 4\,\mu$Jy) exists (Uhuru sources are $\sim 60\%$ of all WS with $S < 4\,\mu$Jy). The incompleteness is great at $|b| > 20°$: at $l = 330$–$30°$ only $\sim 2/3$ of sky area were observed, at $l = 30$–$60°$ this value is $\sim 4/5$ and at $l = 120$–$150° \sim 3/4$, and at $l = 150$–$240° \sim 4/5$. Taking into account the incompleteness (Figure 5a), one may use the deficit of sources at $|l| \approx 120$–$180°$. The deficit of sources at $|l| < 60°$ can be explained partially by confusion effects. However, 2 XRS data at different latitudes (where confusion and incompleteness are little) show that longitude excess at $|l| = 60\text{--}120°$ exists.

(ii) There is a deficit of associations with clusters of galaxies at $l = 240\text{--}360°$ because of the absence of a Palomar Sky Survey at these longitudes.

(iii) The existence of some excess of very weak sources is to be explained by the radiation of galaxies in a local supercluster (in this case $L_x \sim 10^{41}\text{--}10^{42}\,\text{erg s}^{-1}$). The observations of the X-ray background show an excess at $l \sim 130°$ and $\sim 320°$ near the supercluster equator.

If one attempts to interpret the longitude excess as a certain galactic population, the hardness arises because of the absence of sufficient concentration towards the galactic plane. Then the galactic nature of such longitude excess is to be connected with local nonuniformity in the distribution of very close sources of low luminosities.

Other versions are possible, for example:

(i) Novae and related stars or systems of AM Her-type. Then $|\bar{Z}| \sim 0.2\,\text{kpc}$, horizon locates at $\sim 0.3\,\text{kpc}$, $L_x \sim 2 \times 10^{32}\,\text{erg s}^{-1}$. The expected number of sources in the Galaxy is more than 10^6, which contradicts the X-ray background luminosity and the sky distribution of the background.

(ii) Binary systems with condensed stars and normal low mass stars (stellar wind accretion). One would suggest that $|\bar{Z}| \sim 0.3\text{--}0.4\,\text{kpc}$ (as for WS with $S \geqslant 4\,\mu\text{Jy}$), horizon locates at $\sim 0.5\text{--}0.6\,\text{kpc}$, $\bar{L}_x \sim 5 \times 10^{32}\,\text{erg s}^{-1}$. The total number of such sources in the Galaxy is to be more than 7×10^4, which contradicts the rate of collapses in close binary systems and the lifetime of bright X-ray sources (Amnuel and Guseinov, 1980).

(iii) Accretion on a single neutron star with a strong magnetic field. In this case the $|Z|$-distribution would correspond to that of the gas component of the Galaxy, $|\bar{Z}| \sim 0.1\,\text{kpc}$, horizon locates at $\sim 0.15\,\text{kpc}$, $\bar{L}_x \sim 3 \times 10^{31}\,\text{erg s}^{-1}$. For the security of such luminosity a neutron star with a surface magnetic field of more than $10^{10}\,\text{G}$ must move with a velocity of no more than $25\,\text{km s}^{-1}$ in the medium of a density of more than $30\,\text{cm}^{-3}$. Such densities result in giant molecular clouds, which number in the Galaxy ~ 4000 (Solomon and Sanders, 1978). Unfortunately, the value of the molecular clouds number at $R \sim 10\,\text{kpc}$ is not known. Dimensions of a cloud reach $40\,\text{pc}$, then the existence of two clouds at distances $80\text{--}100\,\text{pc}$ from the Sun (longitudes near 90 and 270°) would explain the longitude excess of WS. The expected surface density of the numbers of single neutron stars with necessary parameters is not contradictory to this hypothesis.

Then one would say that the longitude excess of WS ($S < 4\,\mu\text{Jy}$) at $|l| = 60\text{--}120°$ is to be explained partially by the incompleteness of surveys, confusion effects, and the incompleteness of identifications of WS ($S < 4\,\mu\text{Jy}$) with extragalactic objects. Some sources may be single neutron stars accreting interstellar gas in giant molecular clouds, but the reality of such interpretation is uncertain.

4. Conclusions

2 XRS catalog contains data on 677 X-ray sources known as of 1979 September:

Data on 73 bright steady sources ($S > 20\,\mu$Jy) which are galactic in the main.

Data on 159 sources associated with extragalactic objects.

Data on 275 weak steady sources ($S \leqslant 20\,\mu$Jy) unassociated with extragalactic objects.

Data on 63 burst sources (time of bursts from $\sim$s to $\sim$hr).

Data on 40 transient sources.

Data on 24 X-ray Supernova Remnants.

Data on 44 soft sources emitting in the range less than 1 keV (flare stars, systems of RS CVn type, etc.).

Data on 8 hard sources emitting in the range above 20 keV.

Weak steady sources with $S = 4$–$20\,\mu$Jy are, in the main, to be binary systems with normal low mass star and a condensed one (the reason of X-ray emission is in accretion of stellar wind). Weak steady sources with $S < 4\,\mu$Jy are mainly extragalactic.

The total number of bright sources in the Galaxy is ~ 200–$300\ (L_x \gtrsim 10^{36}\,\mathrm{erg\,s}^{-1})$. The population of weak X-ray sources also exists in the Galaxy. The number of such weak sources is $\sim 5 \times 10^3$, $\bar{L}_x \sim 10^{34}\,\mathrm{erg\,s}^{-1}$. A fraction of such sources arises as transient sources and bursters.

Acknowledgements

We acknowledge E. I. Sisoyeva for assistance in the preparation of catalog.

TABLE I
The catalogue

Name XRS	Other names	R.A. decl. (1950)	l b	Area (sq. deg.)	Date		Range keV	Intensity μJy 2–6 keV	Var. comments
(1)	(2)	(3)	(4)	(5)	(6)	(7)	(8)	(9)	(10)
00000+278	A 0000+28	00^{h}00^{m}00^s 27°48′00″	109°56 −32°20	5.600 0	27.9.75 27.9.75	A A	2–18 2–18	90 127	burst burst
00000+726	4U 0000+72	00 00 00 72 36 36	119.30 10.36	0.158 2	1971–73	U	2–6	2.8	−
00012−310	1M 0001−31	00 01 12 −31 03 00	10.85 −78.96	5.100 0	1971–74	O-7	3–10	3.5	−
00058+200	4U 0005+20	00 05 48 20 03 00	109.43 −41.41	0.320 6	1971–73	U	2–6	2.3	
00079+106	0007+106	00 07 57 10 41 48	106.98 −50.63	0.000 001	17–22.8.77	S3	4.14	2.4	
00092−339	4U 0009−33	00 09 12 −33 54 00	353.33 −79.26	4.2079	1971–73 1974–76	U A	2–6 2–18	4 2.8	upper limit
00108+396	4U 0010+39	00 10 48 39 36 00	115.05 −22.41	2.4164	1971–73	U	2–6	2.8	−
00123−052	1M 0012−05	00 12 36 −05 16 30	99.99 −66.24	0.230 0	1971–73	O7	3–10	0.9	−
00153+028	4U 0015+02	00 15 19 02 51 36	106.66 −58.67	1.7139	1971–73	U	2–6	1.07	−
00218+420	1M 0021+42	00 21 48 42 00 00	117.62 −20.32	15.000 0	1971–73	O7	3–10	1.2	−
00224+638	4U 0022+63 Cep X-1, Cep XR-1	00 22 24 63 52 48	120.08 1.43	0.002 2	1971–73 1971–73	U O7	2–6 3–10	12.1 9.7	SNR
00260−097	1M 0026−09	00 26 00 −9 42 00	104.88 −71.49	1.200 0	1971–73	O7	3–10	0.7	−
00262+593	4U 0027+59 A 0026+59	00 26 16 59 19 06	125.91 −3.15	0.003 48	1971–73 22.3–10.4.75 6–7.7.75 12–17.7.75	U A A A	2–6 2.9–7.6 2.9–7.6 2.9–7.6	5.5 60 120 120	− upper limit 3 3
00264−730	4U 0026−73	00 26 24 −73 01 12	305.30 −44.28	0.2175	1971–73	U	2–6	4.07	−
00268−291	4U 0026−29	00 26 53 −29 09 00	11.94 −84.83	1.082 9	1971–73	U	2–6	1.64	−
00288+220	4U 0028+22	00 28 53 22 04 48	116.89 −40.28	0.790 4	1971–73 74–76	U A	2–6 2–18	1.75 3.3	− upper limit
00328+242	1M 0032+24	00 32 48 24 12 00	118.29 −38.25	18	1971–73	O7	3–10	1.1	
00332+588	4U 0033+58	00 33 12 58 51 00	120.95 −03.69	0.206 4	1971–73	U	2–6	2.17	−
00390+411	2U 0039+411 3U 0021+42 4U 0037+39	00 39 00 41 06 00	120.98 −21.46	1.150	1971–73 1974–76	U A	2–6 2–18	1.49 3.0	− −
00397−096	3U 0026−09 2A 0039−096 4U 0037−10	00 39 46 −09 37 48	115.57 −72.10	0.075 0	1971–73 1974–76 1977–78	U A H	2–6 2–18 2–6	4.5 5.3 4.2	− −
00419+368	4U 0041+36	00 41 55 36 49 48	121.43 −25.75	0.744 9	1971–73	U	2–6	8.7	≳3
00421+327	3U 0042+32 2A 0042+323 4U 0042+32	00 42 08.5 32 44 58	121.51 −29.80	0.000 9	1971–73 1971–73 1974–76 11.74 3–6.2.77 14–17.2.77 2.77	U O7 A A A A S3	2–6 3–10 2–18 2–18 2–18 2–18	50 2.1 63 2.5 42 81 81	5 − transient? upper limit transient

Table I (continued)

Name XRS	Other names	R.A. decl. (1950)	l b	Area (sq. deg.)	Date		Range keV	Intensity μJy 2–6 keV	Var. comments
(1)	(2)	(3)	(4)	(5)	(6)	(7)	(8)	(9)	(10)
00500 + 592	1M 0050 + 59 MX 0049 + 50	00^{h}50^{m}00^s 59°13′48″	123°13 −3°37	0.435 0	1971–73	O7	3–10	7.0	–
00503 − 727	SMC X-3	00 50 21.2 −72 42 01	303.15 −44.71	0.000 2	11–16.10.77	S3	2–10	7.5	transient
00520 − 687	4U 0052 − 68	00 52 00 −68 45 00	302.59 −48.65	1.561 0	1971–73	U	2–6	1.57	–
00529 − 739	SMC X-2	00 52 54.5 −73 56 58	303.38 −43.46	0.000 1	11–16.10.77	S3	2–10	10.8	transient?
00536 + 604	MX 0053 + 60 IS 0053 + 604 4U 0054 + 60 MX 0049 + 59	00 53 41.3 60 26 57	123.58 −2.14	0.000 9	1971–73 1971–73 30.12.75– 1.1.76	U O7 S3	2–6 3–10 2–6	7.0 16.5 18.7	– – var?
00549 − 015	2A 0054 − 015 4U 0050 − 01	00 54 55 −01 34 48	126.39 −64.14	0.365 0	1971–73 1974–76	U A	2–6 2–18	2.6 3.7	– –
00554 − 796	1M 0055 − 79	00 55 27 −79 41 13	302.64 −37.71	0.180 0	1971–73	O7	3–10	3.8	–
00577 − 239	1M 0057 − 23	00 57 42 −23 55 30	152.91 −86.01	1.200 0	1971–73	O7	3–10	2.7	
01020 − 242	2A 0102 − 242	01 02 00 −24 15 00	166.60 −85.70	1.000 0	1974–76	A	2–18	1.34	–
01039 − 218	2A 0102 − 222 3U 0057 − 23 4U 0103 − 21	01 03 54 −21 52 30	155.30 −83.52	0.664 4	1971–73 1974–76	U A	2–6 2–18	2.19 1.84	–
01114 − 149	M 0111 − 149	01 11 28.8 −14 55 48	146.87 −76.48	1.110 0	2.1.78	H	2–10	2.49	–
01147 + 650	2S 0114 + 650	01 14 44.3 65 01 34	125.72 −0.01	?	1–3.1.76	S3	2–11	6.35	–
01152 + 634	3U 0115 + 63 4U 0115 + 63	01 15 15.7 63 28 45	125.93 1.02	0.000 2	1.1971 1971–73 12–25.11.74 21.7–16.8.1975 30.12.77 5.1.1978 16.1.78 4.1978	U O7 A A A S3 S3 H	2–6 3–10 2–18 2–18 2–18 2–10 2–10 1–13	120 4.9 6.7 1.3 150 270 470 470	transient – – upper limit transient – – –
01157 − 737	3U 0115 − 73 2A 0116 − 737 4U 0115 − 73 SMC X-1	01 15 46.6 −73 42 06	300.45 −43.58	0.000 09	1971–73 1971–73 1974–76	U O7 A	2–6 3–10 2–18	60 33.4 104	≳ 10 – 30
01206 − 591	2A 0120 − 591 4U 0106 − 59	01 20 38 −59 06 36	295.38 −57.82	0.207 0	1971–73 1974–76	U A	2–6 2–18	3.5 2.3	– –
01209 − 353	2A 0120 − 353 4U 0115 − 36	01 20 55 −35 19 12	264.31 −79.56	0.063	1971–73 1974–76	U A	2–6 2–18	2.3 3.2	– –
01224 + 338	2A 0122 + 338	01 22 29 33 51 00	130.89 −28.24	0.221 0	1974–76	A	2–18	2.7	–
01232 + 075	H 0123 + 075	01 23 14.4 07 34 12	137.56 −54.06	4.200 0	13.1.1978	H	2–10	3.84	var
01296 − 099	4U 0129 − 09	01 29 36 −09 59 24	153.72 −70.14	1.054 2	1971–73	U	2–6	2.19	–
01322 + 007	Cet X-1	01 32 12 +00 45 00	145. −60.	706.5	12.1967	R	1.5–6	950	transient
01345 − 115	4U 0134 − 11	01 34 34 −11 32 24	159.20 −70.87	1.274 8	1971–73 1974–76	U A	2–6 2–18	4.7 0.86	– –
01364 − 182		01 36 25 −18 12 42	133.91 −80.65	optical data	8.1.1975	ANS	0.2–0.28		flare, UV Cet
01385 + 480	4U 0138 + 48	01 38 30 48 03 00	131.48 −13.74	0.278 7	1971–73	U	2–6	1.15	–

Table I (continued)

Name XRS	Other names	R.A. decl. (1950)	l b	Area (sq. deg.)	Date		Range keV	Intensity μJy 2–6 keV	Var. comments
(1)	(2)	(3)	(4)	(5)	(6)	(7)	(8)	(9)	(10)
01427+612	3U 0143+61	01ʰ42ᵐ55ˢ1	129°40	0.000 295	1971–73	U	2–6	7.1	
	4U 0142+61	61°29′59″	−0°40		1971–73	O7	3–10	6.5	
01430−330	U 0143−33	01 43 00	248.83	0.200 0	?				
		−33 00	−76.68						
01486+360	3U 0151+36	01 48 38	136.27	0.501 3	1971–73	U	2–6	3.57	
	4U 0148+36	+36 02 24	−25.02		1974–76	A	2–18	3.3	upper limit
02065−019	H 0206−019	02 06 31.2	162.39	1.530 0	20.1.78	H	2–10	1.25	−
		−01 55 48	−58.48						
02140+045	Cet X-2	02 14	159.	200	26.10.69	R	1.5–5	18700	transient
		04 33	−52.		10.7.70	R	1.5–5	267	upper limit
02140+623	SNR HB3	02 14 00	132.71	optical		H	0.6–2.2	2.91	SNR
	G 132.4+2.2	62 18	1.30	data					
02202+184	H 0220+184	02 20 14.4	151.31	2.21	30.1.78	H	2–10	0.96	−
		18 24 00	−39.14						
02238+312	4U 0223+31	02 23 48	145.70	0.478 4	1971–73	U	2–6	1.27	−
		+31 16 30	−27.12						
02272+437	0227+43	02 27 12	141.16	13	1971–73	O7	3–10	0.9	−
		43 42 00	−15.42						
02285−130	4U 0228−13	02 28 31	186.31	0.808 8	1971–73	U	2–6	3.37	−
		−13 03 36	−62.69		1974–76	A	2–18	2.71	upper limit
02352−526	2A 0235−526	02 35 17	272.24	0.078	1974–76	A	2–18	2.25	−
		−52 37 12	−57.99						
02410+622	3U 0258+60	02 41 01.3	135.64	0.360 7	1971–73	U	2–6	1.99	−
	4U 0241+61	62 15 27	2.24		9–13.7.75	A	300–1200		
					24–29.11.77	S3	2–11	1.4	−
					12.1977	S3	2–11	1.8	
							3.8	22	
02484−853	4U 0248−853	02 48 24	300.29	0.190 6	1971–73	U	2–6	1.62	
		−85 21 00	−31.41						
02522+060	2A 0252+060	02 52 17	169.28	0.192	1974–76	A	2–18	2.25	−
		06 05 24	−45.18						
02528+440	H 0252+440	02 52 50.4	145.32	0.56	14.2.78	H	2–10	1.93	var?
		44 01 48	−13.15						
02530+417	3U 0227+43	02 53 05	146.49	0.062 1	1971–73	U	2–6	5.74	−
	MX 0225+41	41 42 36	−15.18		1971–73	O7	3–10	7.9	1.9
	GX 146−15				1974–76	A	2–18	8.52	−
	2A 0251+413				1975–77	O8	2–20		
	4U 0253+41								
02538+193	H 0253+193	02 53 48	159.28	1.67	6.2.1978	H	2–10	1.02	−
		19 22 12	−34.32						
02554+132	3U 0254+13	02 55 29	164.12	0.041	1971–73	U	2–6	7.46	−
	2A 0255+132	13 13 48	−39.10		1974–76	A	2–18	18.04	−
	4U 0254+13				1975–77	O8	2–20		−
					1971–73	O7	3–10	4.0	−
					14.2.78	H	1–10		−
02586+607	1M 0258+60	02 58 36	138.20	0.23	1971–73	O7	3–10	5.5	−
		60 43 12	1.99						
03029−223	4U 0302−22	03 02 55	210.95	2.617 1	1971–73	U	2–6	1.75	−
		−22 18 00	−59.29						
03049+407		03 04 54	148.98	0.000 87	10.1975	S3	2–6	3.67	β Per
		40 46	−14.90		17–21.8.75	H	0.15–2.7		−
03059+530	1M 0305+53	03 05 55	142.83	0.13	1971–73	O7	3–10	2.1	−
		53 01 12	−4.24						
03064+477		03 06 24	145.59	optical	21–23.9.77	H	0.2–2.8		−
		47 43 42	−8.78	data					
03102+465	4U 0310+46	03 10 12	146.76	0.266 5	1971–73	U	2–6	2.96	−
		46 33 00	−9.45						

Table I (continued)

Name XRS	Other names	R.A. decl. (1950)	l b	Area (sq. deg.)	Date		Range keV	Intensity μJy 2–6 keV	Var. comments
(1)	(2)	(3)	(4)	(5)	(6)	(7)	(8)	(9)	(10)
03110 + 420		03ʰ11ᵐ ˢ 42° ′ ″	149°32 −13°25	18	5.69–4.73	V5	3–12	835–3170	burst
03116 − 227	2A 0311 − 227	03 12 00.9 − 22 46 46	212.89 −57.49	0.000 19	1974–76 12.75–1.76 25–30.1.78	A S3 H	2–18 0.15–0.28 2–10	5.01 1.95	− − −
03118 + 530	3U 0318 + 55 4U 0311 + 53	03 11 50 53 03 00	143.59 −3.76	0.787 2	1971–73 1971–73	U O7	2–6 3–10	2.39 1.9	− −
03129 + 345		03 12 54 34 30 00	153.87 −19.36	optical data	14–15.2.75	ANS	1–3.5	1.73	−
03140 + 380		03 14 00 38 04 42	150.89 −14.59	120		R	0.2–1		burst
03162 − 443	2A 0316 − 443	03 16 12 − 44 18 36	252.75 −56.12	0.210	1974–76 1977–78	A H	2–18 2–6	2.25 0.95	− −
03165 + 413	3U 0316 + 41 2A 0316 + 413 4U 0316 + 41 Per X-1	03 16 27.8 41 19 54.8	150.58 −13.23	0.012 3	1.3.70 1971–73 1974–76 10.75–11.75 1975–77 4–7.1.77 1971–73	R U A S3 O8 S3 O7	0.5–10 2–6 2–18 2–10 2–20 1.5–15.5 3–10	250 79 83 17.3 2.73 6.18 36.8 50.3	− − − extended source point source − − −
03210 − 450	3U 0302 − 47 4U 0321 − 45	03 21 00 − 45 01 30	253.51 −55.09	0.621 5	1971–73 1974–76	U A	2–6 2–18	3.04 1.25	− upper limit
03210 + 236		03 21 23 36 36	162.36 −27.18	120			0.2–1		burst
03226 + 595	4U 0322 + 59	03 22 36 59 33 00	141.37 2.53	0.214 4	1971–73	U	2–6	3.42	−
03228 + 285	H 0324 + 28	03 22 48 28 30 36	159.43 −23.01	0.1	17–22.8.77	H	0.15–2.8	10.6	
03250 + 440		03 25 44	150. −10.	wide region	6.7.1974	B	20–210	3000	burst
03277 + 432	A 0327 + 43	03 27 48 43 49 48	150.91 −10.03	0.14	2.6.78 2.1977 16.9–31.7.78	A A A	2–18 2–18 2–18	10.6 2.12 15.78	− upper limit −
03318 − 363	A 0331 − 36	03 31 48 − 36 18 00	237.94 −54.58	optical data	1974–76 26.1.78	A H	2–18 2–10	1.3 3.2	− upper limit
03330 + 317	H 0333 + 317	03 33 04.8 31 46 48	159.18 −19.09	1.16	18.2.1978	H	2–10	1.02	−
03342 − 302	4U 0334 − 30	03 34 12 − 30 12 00	227.52 −53.90	0.871 3	1971–73	U	2–6	1.47	−
03342 + 002		03 34 13.2 00 25 33	184.91 −41.57	optical data	17.8.77 3–10.2.78	H C	0.2–2.8 2.5–7.5		soft −
03353 + 096	2A 0335 + 096 4U 0344 + 11	03 35 19 09 37 48	176 29 −35.29	0.414 0	1971–73 1974–76	U A	2–6 2–18	3.41 3.67	− −
03362 + 010	4U 0336 + 01	03 36 12 01 01 12	184.70 −40.81	1.584 4	1971–73	U	2–6	167	transient
03385 + 500	A 0338 + 50	03 38 34 + 50 03 00	148.67 −30.94	1.512 0	11.1974 7–8.75	A A	1.2–5.8 2.4–19.8	1.84 4.51	− −
03432 − 536	3U 0328 − 52 2A 0343 − 536 4U 0339 − 54	03 43 17 − 53 37 48	264.68 −48.76	0.110 0	1971–73 1971–73 1974–76	U O7 A	2–6 3–10 2–18	2.81 2.1 6.35	− − 10
03492 − 139	2A 0349 − 139	03 49 12 − 13 56 24	204.58 −46.02	0.296 0	1974–76	A	2–18	1.34	−
03522 + 308	3U 0352 + 30 2A 0352 + 309 4U 0352 + 30	03 52 14.4 30 53 42	163.09 −17.11	0.000 87	1971–73 1971–73 10.72–1.75 1974–76	U O7 C A	2–6 3–10 2.5–7.5 2–18	50.1 3.4 54.7 50.9	3 − 2 1.7

Table I (continued)

Name XRS	Other names	R.A. decl. (1950)	l b	Area (sq. deg.)	Date		Range keV	Intensity µJy 2–6 keV	Var. comments
(1)	(2)	(3)	(4)	(5)	(6)	(7)	(8)	(9)	(10)
					1976	B	20–150		
					21.02–2.3.76	O8	2–60	29.5	–
					1977	S3	2–6	32.5	–
03531−400	A 0353−40	03ʰ53ᵐ09ˢ −40°01′00″	243°34 −50°37	0.250 0	12.3.75	A	2–18	103	burst
					11.10.75	A	2–18	130	burst
03576−743	4U 0357−74	03 57 36 −74 19 30	288.58 −37.32	0.190 2	1971–73	U	2–6	2.02	–
04040+476	4U 0404+47	04 04 00 +47 40 30	153.46 −3.09	0.202 0	1971–73	U	2–6	1.42	–
04064−308	4U 04067−308	04 06 24 −30 52 30	229.94 −47.10	3.4719	1971–73	U	2–6	1.03	–
04074+379	4U 0407+37	04 07 24 37 55 30	160.57 −9.85	0.334 5	1971–73	U	2–6	1.89	–
04106+103	3U 0405+10 2A 0411+103 4U 0410+10	04 10 40.4 10 20 37.5	182.49 −28.15	0.000 008	1971–73 1971–73 1974–76 1977–78	U O7 A H	2–6 3–10 2–18 2–6	4.46 2.8 5.45 1.92	– –
04150−120		04 15 00 −12 00 00	205. −40.	706	1.11.72	R	0.18–2		soft, SNR
04150+379	H 0415+37	04 15 01.0 37 54 20.0	161.68 −8.83	optical data	1977–78	H	2–10	2.02	
04215+347	3U 0430+37 4U 0421+34	04 21 30 34 43 30	164.90 −10.12	0.264 6	1971–73 1971–73	U O7	2–6 3–10	2.72 0.9	– –
04234−531	3U 0400−59 4U 0423−53	04 23 24 −53 09 00	261.55 −43.17	0.232 8	1971–73 1971–73 1974–76	U O7 A	2–6 3–10 2–18	2.99 0.8 1.5	– – upper limit
04276−077	4U 0427−07	04 27 36 −07 42 00	202.84 −34.80	1.824 2	1971–73 1974–76	U A	2–6 2–18	3.37 1.9	– upper limit
04293−310	4U 0429−31	04 29 22 −31 00 00	231.23 −42.26	3.295 7	1971–73	U	2–6	1.5	–
04305+052	4U 0432+05 2S 0430+05	04 30 32.7 05 14 06.2	190.10 −26.65	0.000 013	1971–73 1974–76 16–25.11.75	U A S3	2–6 2–18 2–10	3.34 5.3 5.3	– – –
04309−615	3U 0426−63 2A 0430−615 4U 0427−61	04 30 58 −61 34 12	272.14 −40.10	0.077 0	1971–72 1971–73 1974–76	U O7 A	2–6 3–10 2–18	3.49 2.5 9.0	– – 4
04312−136	2A 0431−136 4U 0431−12	04 31 12 −13 36 36	209.85 −36.61	0.067 0	1971–73 1974–76 1977–78	U A H	2–6 2–6 2–6	3.8 4.0 1.64	– –
04360+120		04 36 00 12 00 00	185.19 −22.36	113	1971	O6	27–189		hard burst
04400+069	1M 0440+46	04 40 02 06 59 24	190.27 −24.46	0.500 0	1971–73	O7	3–10	2.8	–
04432−095	3U 0431−10 4U 0443−09	04 43 12 −9 30 00	206.88 −32.20	1.151 2	1971–73 1971–73 1974–76	U O7 A	2–6 3–10 2–18	2.74 2.8 3.7	– – upper limit
04461+449	4U 0446+44	04 46 07 44 57 36	160.47 0.24	0.007 4	1971–73 1971–73 1975–77 1978	U O7 O8 H	2–6 3–10 2–20 2–11	8.40 6.9 4.17	– –
04476−037	H 0447−037	04 47 40.8 −3 45 36	201.60 −28.52	1.490 0	3.9.1977	H	2–10	1.60	–
04490−550	H 0449−55	04 49 00 −55 00 00	263. −39.		2.1978	H	2–11 2–11	140 5	burst upper limit
04495+668	1M 0449+66	04 49 31 66 50 24	143.62 14.43	0.270 0	1971–73	O7	3–10	7.1	–
04527−742	H 0452−742	04 52 45.6 −74 13 48	286.34 −34.06	2.04	23.10.77	H	2–10	0.93	–

Table I (continued)

Name KRS	Other names	R.A. decl. (1950)	l b	Area (sq. deg.)	Date		Range keV	Intensity μJy 2–6 keV	Var. comments
1)	(2)	(3)	(4)	(5)	(6)	(7)	(8)	(9)	(10)
04566−449	3U 0510−44	04ʰ56ᵐ41ˢ	250°.17	1.130	12.70–3.71	U	2–6	3.34	−
	2A 0456−449	−44°55′48″	−35°.38		1974–76	A	2–18	3.17	−
04574−357	4U 0457−35	04 57 24	238.56	2.090 3	1971–73	U	2–6	0.40	−
		−35 42 00	−37.29						
05000−554		05 00 00	263.38	30	9.2.1978	H	2–20	217	burst
		−55 26 00	−36.44						
05032+357	A 0503+35	05 03 12	169.69	0.897	12–25.11.74	A	2–6	7.2	−
		+35 45 00	−2.98		21.7–16.8.75	A	2–18	5.8	−
05048−843	4U 0504−84	05 04 48	297.18	0.316 1	1971–73	U˙	2–6	1.2	−
		−84 22 30	−29.75						
05050−213	4U 0505−21	05 05 00	222.18	1.958 3	1971–73	U	2–6	1.2	−
		−21 21 00	−32.01		1974–76	A	2–18	3.2	−
05065−034	4U 0506−03	05 06 30	203.81	0.248 6	1971–73	U	2–6	2.3	−
		−03 25 30	−24.25						
05096+019	4U 0509+01	05 09 36	199.16	2.178 3	1971–73	U	2–6	2.99	−
		01 57 00	−20.93						
05106−446	1M 0510−44	05 10 38	250.03	0.27	1971–73	O7	3–10	0.5	−
		−44 39 36	−35.89						
05124−400	MX 0513−40	05 12 29	244.36	0.000 097	1971–73	O7	1–10	21.4	5
					1971–73	U	2–6	30	3
	2A 0512−399	−40 05 53	−34.99		19.9.72	U	2–6	265	burst
	4U 0513−40				1974–76	A	2–18	18.5	8
	MXB 0512−40				12.1976	S3	2–11	4	
					7–8.77	S3	1–10	14.1	var, burst
05130+459		05 13 00	162.59	1.8	5.4.74	R	0.2–1.6		soft
		45 57	4.57		14–16.8.77	H	0.2–2.8		−
05150+384	4U 0515+38	05 15 02	168.95	0.446 1	1971–73	U	2–6	1.24	−
		34 24 00	0.49						
05174−456	H 05174−456	05 17 26.4	251.38	2.55	3.9.1977	H	2–10	0.73	−
		−45 39 36	−34.79						
05175+175	4U 0517+17	05 17 34	186.48	0.955 6	1971–73	U	2–6	1.2	−
		17 34 12	−10.95						
05185−262	4U 0518−26	05 18 34	228.73	0.383 1	1971–73	U	2–6	6.59	−
		−26 12 00	−30.63						
05198+065	4U 0519+06	05 19 48	196.37	2.375 3	1971–73	U	2–6	1.03	−
		06 32 24	−16.40		1974–76	A	2–18	2.93	upper limit
05212−365		05 21 14	240.60	radio		H	3–6	1	
		−36 30 00	−32.71	data					
05213−719	LMC X-2	05 21 16.5	283.1	0.000 07	1971–73	U	2–6	27	2
	3U 0521−72	−72 00 19	−32.7		1971–73	O7	3–10	22.4	
	2A 0521−720				1974–76	A	2–18	31.7	−
	4U 0520−72				8–9.77	H	0.9–13.3	16	
					17.8.1977– 10.2.1978	H	1.5–13.5		
05263−683	LMC X-5	05 26 19	278.77		1971–73	O7	3–10	5.3	−
	MX 0528−68	−68 23 24	−32.91	0.001 8					
05263−661	FXP 0520−66	05 26 18	276.08	?	5.3.1979			50	hard
		−66 07 00	−33.21						burst
05265−350	U 0527−35	05 26 30	239.93	0.44	1971–73	U	2–6		−
		−35.41	−31.50						
05276−328	2A 0526−328	05 27 37.1	236.81	0.000 35	1974–76	A	2–18	3.7	−
		−32 52 59	−30.61		9.77, 10.78	H	0.1–3		−
					7–13.9.77	H	2–6	3.3	−
					5–12.3.78	H	2–6	4.3	−
05303−370	3U 0530−37	05 30 19	241.63	1.600 0	12.70–4.71	U	2–6	5.2	
		−37 00 00	−31.04		1971–73	O7	3–10	1.6	
					1974–76	A˙	2–18	4.0	upper limit

Table I (continued)

Name XRS	Other names	R.A. decl. (1950)	*l* *b*	Area (sq. deg.)	Date		Range keV	Intensity μJy 2–6 keV	Var. comments
(1)	(2)	(3)	(4)	(5)	(6)	(7)	(8)	(9)	(10)
05315+219	Tau X-1 Tau -1 4U 0531+21	05^{h}31^{m}30^s 21°58′52″	184°56 −5°79	0.0002	1971–73 1971–73	U O7	2–6 3–10	1581 1058.8	SNR −
05328−663	LMC X-4 4U 0532−66 2A 0532−664	05 32 51.7 −66 23 44	276.60 −32.55	0.00087	1971–73 1971–73 1974–76 20–26.2.76	U O7 A S3	2–6 2–6 2–18 2–6	2.62 4.7 9.0 36.9	− upper limit − 6
05328−056	3U 0527−05 2A 0532−056 4U 0531−05	05 32 49.2 −05 25 08	209.0 −19.4	0.0002	1971–73 1971–73 1974–76 9.77, 1.78	U O7 A S3	2–6 3–10 2–18 2–11	5.21 4.8 8.5 1.3	
05347−581	H 0534−581	05 34 45.6 −58 07 36	266.49 −32.69	0.52	27.2.77	H	2–10	1.05	−
05357−668	A 0538−66	05 35 42.5 −66 52 40	275.58 −32.29	0.00054	6–7.77 17.8.77 9.10.77	A H H	2–17 1.5–13.5 1.5–13.5	37 18	burst −
05357+262	A 0535+26 4U 0538+26	05 35 47 26 16 52	181.47 −2.6	0.0004	1971–73 28.4.75 20.9.75 7.11.75 14.11.1975 12–15.5.77	O7 A S3 S3 S3 S3	3–10 2–18 1–10 1–10 1–10 2–20	6.2 3000 12 317 80 317	upper limit upper limit
05370−441	PKS 0537−441	05 37 00 −44 06 00	250.06 −31.00	radio data		H	3–6	3.1	
05382−661	LMC bar	05 38 16 −66 09 55	276.00 −32.00	optical data	6.11.73	R	0.4–1.5		soft
05389−641	LMC X-3 3U 0539−64 2A 0539−642 4U 0538−64	05 38 38.1 −64 06 20	274.11 −32.09	0.001	1971–73 1971–73 1974–76 8–11.77 17.8.77– 10.2.78	U O7 A H H	2–6 2–6 2–18 0.9–13.3 1.5–13.5	42 18.7 27.1 5.0	3 − − − −
05390−669	LMC X-4?	05 39 00 −66 54 00	276.86 −31.88	0.8	30.6–16.7.77	A	2–17	250	burst
05401−697	LMC X-1 3U 0540−69 2A 0540−698 4U 0540−69	05 40 04.8 −69 46 09	280.23 −31.44	0.000055	1971–73 1971–73 1974–76 2.1976 8.77–2.78	U O7 A S3 H	2–6 3–10 2–18 1.5–6 1.5–13.5	33.4 16.2 27.1 27.7	1.5 − − 1.5
05418+608	4U 0541+60	05 41 48 60 51 00	152.03 16.08	0.3690	1971–73	U	2–6	1.54	−
05438−316	3U 0545−32 4U 0543−31	05 43 48 −31 39 00	236.54 −26.97	0.3794	1971–73 1971–73 12.70–4.73 1974–76	U O7 U A	2–6 3–10 2–6 2–18	1.87 3.2 6.7 3.3	− − − upper limit
05441−665	LMC? H 0544−665	05 44 11.5 −66 35 24	276.5 −31.4	0.00091	17.77– 10.2.1978	H	2–11 1.5–13.5	1.8 1.8	−
05464−882	4U 0546−88	05 46 24 −88 13 30	301.07 −27.87	0.1858	1971–73	U	2–6	2.76	−
05480+290	4U 0548+29	05 48 00 29 00 00	180.56 1.08	3.1461	1971–73	U	2–6	4.41	−
05488−322		05 48 48.9 −32 17 07.4	235.66 −25.60		9.77, 3./8 19–21.9.77 16–18.3.78	H H	3–6 0.15–3	3 1.82	−
05497−074		05 49 45.8 −07 27 55	212.68 −16.44	0.00038	17–23.1.78 19.9.1977	S3 H	2–11	2.5	− −
05512+466	2A 0551+466	05 51 15	165.57	0.093	1971–73	U	2–6	3.32	−

Table I (continued)

Name XRS	Other names	R.A. decl. (1950)	*l* *b*	Area (sq. deg.)	Date		Range keV	Intensity μJy 2–6 keV	Var. comments
(1)	(2)	(3)	(4)	(5)	(6)	(7)	(8)	(9)	(10)
	4U 0558+46	ʰ ᵐ ˢ 46°37′12″	10°.51 °.		1974–76 8.77–2.78	A H	2–18 2–10	7.2	10
05538−486	4U 0553−48	05 53 53 −48 40 30	255.82 −29.03	0.311 9	1971–73	U	2–6	2.19	−
05570−381	4U 0557−38	05 57 00 −38 06 00	244.31 −26.18	1.299 4	1971–73	U	2–6	1.40	−
05598−571	4U 0559−57	05 59 48 −57 10 30	265.61 −29.29	0.315 6	1971–73	U	2–6	0.7	−
06002+465	MX 0600+46	06 00 12 46 30 00	166.41 11.85	1.142	1971–73 29.9.71– 18.5.74	O7 O7	3–10 2–6 3–10	6.0 7.8 5.3	var? var?
06084−491	4U 0608−49	06 08 24 −49 09 00	256.85 −26.77	0.533 9	1971–73	U	2–6	2.02	−
06088+497	2A 0608+497 MX 0600+46(?)	06 08 50 49 43 48	164.11 14.54	0.344	1974–76	A	2–18	2.3	−
06137+224	3U 0620+23 4U 0617+23 1M 0614+22	03 13 45 22 29 00	189.10 2.88	0.033 3	1971–73 1971–73 24.11.74 9.1975	U O7 R A	2–6 3–10 0.75–3 1.1–7.5	2.05 3.2	SNR −
06143+091	4U 0614+09	06 14 22.7 09 09 08	200.85 −3.39	0.000 1	24.9.1970 1971–73 1971–73 2.1973 5.1975 8.1975 9.1975 1975–77	R U O7 C A A O8 S3	1.2–16 2–6 3–10 2–7 2–7 2–7 2–11	107.5 200 29.7 78 75 150 47	5 1.7 burst
06148+153	4U 0614+15	06 14 48 15 18 00	195.53 −0.34	1.927 7	1971–73	U	2–6	9.2	4
06150+093	MX 0615+09 3U 0614+09(?)	06 15 +09 18	200.83 −3.16	0.125 6	12.2.1976	S3	1.3–5	534	burst
06201−003	A 0620−00	06 20 11.176 −00 19 10.80	210.19 −6.53	optical data	1971–73 3.8.1975 14.8.1975 27.8.1975 10.9.1975 14.9.1975 7.1.1976 2.1976 3.1976	O7 A A S3 Sal 4 S3 S3 A A	3–10 2–18 2–18 1.5–5 2–10 0.6–0.9 1.5–5 1.5–5 3–6 3–6	6.8 70 83500 25300 18900 11900 300 1500 70	upper limit transient maximum − − − − − second maximum upper limit
06216+117	4U 0621+11	06 21 38 11 46 48	199.41 −0.55	3.555 9	1971–73	U	2–6	2.79	−
06225−529	H 0622−529	06 22 30 −52 57 36	260.18 −25.09		6–9.10.77	H	0.4–3		−
06261−541	3U 0624−55 2A 0626−541 4U 0627−54	06 26 10 −54 10 12	262.95 −25.14	0.162	1971–73 1971–73 1974–76 1975–77	U O7 A O8	2–6 3–10 2–18 2–20	4.39 1.2 3.7	− − − −
06270+216		06 27 00 21 37 48	191.30 5.19		1974–76	A	2–18	1.40	−
06272+675	4U 0627+67	06 27 14 67 34 12	147.40 23.16	2.885 8	1971–73	U	2–6	1.40	−
06273−381	4U 0627−38	06 27 18 −38 06 00	246.25 −20.49	0.230 0	1971–73	U	2–6	7.8	6
06288−284	4U 0628−28	06 28 48 −28 24 00	236.81 −16.72	2.396 4	1971–73	U	2–6	4.61	−

Table I (continued)

Name XRS	Other names	R.A. decl. (1950)	l b	Area (sq. deg.)	Date	(7)	Range keV	Intensity μJy 2–6 keV	Var. comments
(1)	(2)	(3)	(4)	(5)	(6)	(7)	(8)	(9)	(10)
)6300 + 024	4U 0630 + 02	06^{h}30^{m}00^s 02°24′00″	208°68 −3°11	1.361 8	1971–73	U	2–6	2.5	–
)6350 − 033	4U 0635 − 03	06 35 00 −03 19 30	214.35 −4.63	3.175 5	1971–73	U	2–6	3.52	–
)6384 + 742	4U 0638 + 74	06 38 24 74 12 00	140.55 25.57	5.956 1	1971–73 1974–76	U A	2–6 2–18	2.3 2.7	– upper limit
)6429 − 166		06 42 54 −16 39	227.79 −8.89	optical data	3–5.6.75 4–9.10.77	ANS H	0.2 0.15–2.5		– –
)6430 + 534	H 0643 + 534	06 43 04.8 53 29 24	162.53 20.87	0.96	1.10.1977	H	2–10	0.96	–
)6560 − 071	MX 0656 − 07	06 56 01 −07 11 42	220.17 −1.66	0.007 8	1971–73 1971–73 20.9.1975 4.10.1975 19.3.1976 27.3.1976	O7 U S3 S3 S3 S3	3–10 2–6 1.3–13 1.3–13 1.3–13 1.3–13	4.5 127 32 80 110	upper limit – – – –
)6560 − 031	4U 0656 − 03	06 56 00 −03 07 12	216.57 0.12	1.893 5	1971–73	U	2–6	1.40	–
)6572 − 114	2A 0709 − 114	06 57 16.8 −11 28 12	224.14 −3.45	0.310	1974–76	A	2–18	7.2	–
)6576 − 351		06 57 36 −35 06 00	245.69 −13.73	2.2	1971–73	O7	3–10	0.5	–
)7002 − 563	H 0700 − 563	07 00 12 −56 22 48	266.66 −21.05	0.59	5.11.77	H	2–10	1.09	–
)7056 + 186		07 05 40 18 37 06	194.56 4.92	1.4	29.9–4.10.77 26–31.3.1978	H	2–6	0.7	–
)7080 − 357	3U 0657 − 35 2A 0708 − 357	07 08 00 −35 42 00	247.15 −12.05	1.070	12.70–3.1971 1974–76	U A	2.4–6.9 2–18	8.5 2.7	– –
)7083 − 168	4U 0708 − 16	07 08 22 −16 49 12	230.14 −3.52	0.821 7	1971–73	U	2–6	1.28	–
)7084 − 492	4U 0708 − 49	07 08 24 −49 15 00	260.02 −17.44	0.333 8	1971–73	U	2–6	3.66	–
)7090 − 221	2A 0709 − 221	07 09 00 −22 10 48	234.98 −5.86	0.422	1974–76	A	2–18	7.2	–
)7107 + 456	2A 0710 + 456	07 10 46 45 37 48	172.05 22.82	0.150	1974–76	A	2–18	2.7	–
)7113 − 384	4U 0711 − 38	07 11 18 −38 24 00	249.94 −12.61	0.641 9	1971–73	U	2–6	1.23	–
)7124 − 113	H 0712 − 113	07 12 28.8 −11 21 00	225.76 −0.08	0.89	15.10.77	H	2–10	1.26	–
)7183 − 546	3U 0705 − 55 4U 0718 − 54	07 18 22 −54 36 00	265.83 −18.02	1.070 7	1971–73 1974–76	U A	2–6 2–18	7.0 2.7	4 upper limit
)7202 + 558	4U 0720 + 55	07 20 12 55 52 48	161.45 26.65	6.570 5	1971–73 1974–76	U A	2–6 2–18	3.62 1.8	– –
)7267 − 260	4U 0728 − 25 2A 0726 − 260	07 26 43.2 −26 04 48	240.33 −4.11	0.02	1971–73	U	2–6	2.87	–
)7285 + 060		07 28 27.12 06 05 06	212.09 11.53	optical data	1974–76	A	2–18	1.2	–
)7290 − 379	4U 0729 − 37	07 29 00 −37 54 00	251.06 −9.27	2.489 6	1971–73	U	2–6	0.77	–
)7333 − 186	4U 0733 − 18	07 33 22 −18 36 00	234.54 0.86	1.099 6	1971–73	U	2–6	1.12	–
)7360 − 500		07 36 00 −50 00 00	263. −14.	530	1.3.76	S3	1.3–5		burst
)7373 − 108	4U 0737 − 10	07 37 19 −10 48 36	228.21 5.52	1.382 9	1971–73	U	2–6	3.77	–

Table I (continued)

Name XRS	Other names	R.A. decl. (1950)	l b	Area (sq. deg.)	Date		Range keV	Intensity μJy 2–6 keV	Var. comments
(1)	(2)	(3)	(4)	(5)	(6)	(7)	(8)	(9)	(10)
07380+498	2A 0738+498	$07^h38^m02^s$ 49°53′24″	168°62 28°26	0.584	1974–76	A	2–18	2.7	–
07390−199	4U 0739−19 2A 0745−191	07 44 52.8 −19 07 12	236.34 2.97	0.038	1971–73	U	2–6	2.74	–
07394+036		07 39 28 03 41 06	215.67 13.11	optical data	19.10.74	ANS	0.2–0.28		soft, YZ CMi
07401+290		07 40 11.4 29 00 22	191.19 23.27		14–16.10.77	H	0.2–2.8		–
07429−286	3U 0757−26 4U 0742−28	07 57 48 −26 24 00	244.12 −1.75	0.9	1971–73 1971–73	U O7	2–6 3–10	0.3 1.1	– –
07517+222	H 0751+222	07 51 46 22 13 48	199.09 23.35	0.79	20.10.77	H	2–10	1.26	–
07521+221		07 52 08 22 08 17	199.92 23.39	optical data	17–22.10.77 10.76, 4.77	H O8	0.15–0.5 2–8	0.04 1.7	burst, U Gem
07574−484	3U 0750−49 4U 0750−49	07 57 27 −48 26 48	262.96 −9.85	0.023 1	1971–73 1971–73 5.1974 10–25.11.74	O7 U C A	3–10 2–6 2.5–20	1.4 1.3 1.5 1.3	upper limit
07578−264		07 57 48 −26 24 00	244.12 1.75	0.9	1971–73	O7	3–10	1.1	–
08048−530		08 04 48 −53 03 00	267.58 −11.21	1.0	1971–73	O7	3–10	1.4	–
08081−352		08 08 06 −35 13 00	252.76 −1.15		1979	H	0.15–0.5 0.5–2		–
08134−385	4U 0813−38	08 13 24 −38 33 00	256.10 −2.11	0.158 6	1971–73	U	2–6	1.0	–
08142−567	A 0813−57 4U 0814−56	08 14 12 −56 45 00	271.52 −12.01	0.261 2	1971–73 11.1974	U A	2–6 2.5–20	2.40 3.7	– 2.7
08152−075	2A 0815−075	08 15 12 −07 30 00	230.09 15.26	0.450	1974–76	A	2–18	3.7	–
08215−427	3U 0821−42 4U 0821−42	08 21 35 −42 44 42	260.37 −3.21	0.010 9	1971–73 1971–73 11.1974	U O7 A	2–6 3–10 2–18	7.73 4.0 11.3	
08336−450	4U 0833−45 Vel X-1 Vel X-2	08 33 36 −45 03 00	263.58 −2.82	0.052	1971–73 1971–73 10–25.11.74 23.7.1976	U O7 A A	2–6 3–10 2–18 2–18	13.24 25.5 27.0 74.5	SNR 1.5 burst
08356−483	A 0835−48 4U 0842−47	08 35 36 −48 21 00	266.43 −4.54	0.21	1971–73 10–25.11.74	U A	2–6 2.5–20	4.38 9.0	– –
08362−426	MX 0836−42 4U 0836−42	08 36 12 −42 36 54	261.93 −0.97	0.012 1	1971–73 1971–73 12–25.11.74 21.7–16.8.75 9–12.1.76	U O7 A A A	2–6 3–10 2–18 2–18 2–18	78 24.3 1.6 0.8 1.0	transient upper limit upper limit upper limit
08429−349	4U 0842−34	08 42 58 −34 55 12	256.68 4.81	1.411 3	1971–73	U	2–6	5.8	5
08449−531	H 0844−531	08 44 58 −53 08 24	271.14 −6.32	3.000 0	9.12.1977	H	2–10	1.83	var?
08450−296	4U 0845−29	08 45 00 −29 40 30	285.81 8.40	1.095 2	1971–73	U	2–6	6.0	2
08514−469	4U 0842−47 A 0854−46 2A 0851−469	08 51 26.4 −46 55 12	266.98 −1.57	0.051 0	1974–76	A	2–18	7.2	–
08542−445	4U 0854−44	08 54 12 −44 30 00	265.45 0.37	0.393 5	1971–73 10–25.11.74 21.7–16.8.75	U A A	2–6 2–18 2–18	6.61 6.7 2.27	– – –

Table I (continued)

Name XRS	Other names	R.A. decl. (1950)	l b	Area (sq. deg.)	Date		Range keV	Intensity μJy 2–6 keV	Var. comments
(1)	(2)	(3)	(4)	(5)	(6)	(7)	(8)	(9)	(10)
08590＋509	2A 0859＋509	08ʰ59ᵐ02ˢ 50°57′00″	167°87 41°15	1.448 0	1974–76	A	2–18	1.3	–
09002－403	3U 0900－40 Vel X-1 GX 263＋3	09 00 13.25 －40 21 25.2	263.07 3.93	optical data	17.5.1967 2.2.1968 13.5.1970 1971－73 1971－73 10–25.11.74	R R R U O7 A	2–16 4–11 1.4–20 2–6 3–10 2–18	430 580 70 410 47.5 180	– – – 10 – 4
09020＋573		09 02 00 57 18 00	150. 7.	113	1.9.1969	V5	3–12	5840	burst
09062－095	3U 0901－09 2A 0906－095 4U 0900－09	09 06 17 －09 32 24	239.33 24.67	0.085	1971–73 1971–73 1974–76	U O7 A	2–6 3–10 2–18	6.95 2.1 7.2	– – –
09084－669	4U 0908－66	09 08 24 －66 55 30	283.73 －12.98	0.485 4	1971–73	U	2–6	20.0	–
09132－461	4U 0913－46	09 13 12 －46 09 00	268.92 1.72	1.790 2	1971–73	U	2–6	1.10	–
09177＋634	3U 0917＋63	09 17 45 63 27 00	150.99 40.66	0.190	12.70-3.71 1971–73 1974–76	U O7 A	2–6 3–10 2–18	6.7 2.0 2.3	var? upper limit
09191－549	4U 0919－54 3U 0918－55	09 19 06 －54 57 00	275.84 －3.79	0.010 1	1971–73 1971–73 10–25.11.74	U O7 A	2–6 3–10 2.5–20	9.15 12.1 27.0	– – 6
09207－628	2A 0920－628	09 20 43.2 －62 49 48	281.60 －9.22	0.115 0	1974–76	A	2–18	9.9	–
09214－630	2S 0921－630 2A 0920－628 H 0921－631	09 21 25.4 －63 04 27	281.83 －9.33	0.000 22	1974–76 10–14.2.78 6.1.1978	A S3 H	2–18 5.2 2–10	9.9 2.3 2.0	– – –
09233－314	A 0921－31 2A 0922－317 4U 0923－31	09 23 20 －31 29 06	259.74 13.41	0.051 3	1971–73 1974–76	U A H	2–6 2–18 2–10	4.44 5.8 8.2	– – 2
09433－140	2A 0943－140 4U 0937－12	09 43 22 －14 01 48	249.66 28.84	0.065 0	1971–73 1974–76 25.11.78	U A H	2–6 2–18 1–13	1.40 4.3	– – –
09436＋712	1M 0943＋71	09 43 36 71 15 36	140.47 39.25	5.300 0	1971–73	O7	3–10	4.0	–
09459－306	3U 0946－30 2A 0946－310 4U 0945－30	09 45 54 －30 40 30	262.79 17.33	0.087 8	1971–73 1971–73 1974–76 19–23.2.77	U O7 A S3	2–6 3–10 2–18 2–10	2.89 5.1 5.0 4.5	– – – –
09483＋121		09 48 19.9 12 06 43.2	215.54 31.89	optical data	1974–76	A	2–18	1.21	–
09544＋700	3U 0953＋71 2A 0954＋700 4U 0954＋70	09 54 24 70 02 24	141.10 40.68	0.311 0	1971–73 1974–76 22.10.77 19.4.78	U A H	2–6 2–18	3.99 3.2	– – –
09555－284	4U 0955－28	09 55 31 －28 24 36	262.88 20.41	1.356 8	1971–73	U	2–6	0.78	–
09587－359	U 0957－37	09 58 42 －35 56 00	267.57 15.88	0.760 0	1971–73	U	2–6		
10088＋138	A 1020＋13	10 08 48 13 48 00	201.22 50.33	1.200 0	1974–76	A	2–18	2.52	–
10144－579	A 1014－57	10 14 24 －57 57 00	283.64 －1.27	0.120 0	10–25.11.74	A	2.5–20	13.5	6
10151－254	4U 1015－254	10 15 06 －25 24 00	264.49 25.50	2.404 7	1971–73 1974–76	U A	2–6 2–18	8.8 1.35	3 upper limit
10184＋498	A 1018＋49	10 18 24 49 48 00	164.68 53.62	1.000 0	1974–76	A	2–18	1.89	–

Table I (continued)

Name XRS	Other names	R.A. decl. (1950)	l b	Area (sq. deg.)	Date		Range keV	Intensity μJy 2–6 keV	Var. comments
1)	(2)	(3)	(4)	(5)	(6)	(7)	(8)	(9)	(10)
10220−408	4U 1022−40	10ʰ22ᵐ02ˢ −40°51′00″	275°24 13°72	0.3727	1971–73	U	2–6	11.7	6
10224−554	3U 1022−55	10 22 29 −55 29 14	283.24 1.40	0.0810	12.70–3.71	U	2–6	17.4	−
					1971–73	O7	3–10	3.9	−
					12–25.11.74	A	2–18	4.0	upper limit
					21.7–16.8.75	A	2–18	1.3	upper limit
					9–12.1.76	A	2–18	5.3	upper limit
10270+590		10 27 00 59 00 00	150.77 50.05	great region	1976	R	0.1–0.4		soft, burst
10284+512	H 1028+512	10 28 26 51 17 24	161.20 54.41	2.2300	10.11.77	H	2–10	0.94	−
10335−270	3U 1044−30 2A 1033−270 4U 1033−26	10 33 31 −27 00 36	269.27 26.60	0.1820	1971–73	U	2–6	2.29	−
					1971–73	O7	3–10	5.6	upper limit
					1974–76	A	2–18	3.2	−
					1975–77	O8	2–20		−
10346−565	A 1034−56 4U 1037−60 2A 1028−571	10 34 36 −56 33 00	285.23 1.35	0.0260	1971–73	U	2–6	3.77	7
					10–25.11.74	A	2–18	13.5	2
					1974–76	A	2–18	24.2	7
10413−079	2A 1041−079	10 41 19 −07 55 48	257.21 43.07	0.3900	1974–76	A	2–18	1.8	−
10416−218	4U 1041−21	10 41 36 −21 48 00	267.73 31.96	1.1399	1971–73	U	2–6	11.8	4
10440−594	A 1044−59 4U 1053−58	10 44 00 −59 27 00	287.71 −0.60	0.0630	13.5.70	R	0.2–2.7		
					1971–73	U	2–6	3.77	
					10–25.11.74	A	2–18	9.0	2
					17–19.7.75	O8	2–20		
10450−593	4U 1043−59	10 45 00 −59 23 00	287.75 −0.40	radio data	1971–73	U	2–6	3.36	SNR
					17–19.7.75	O8	2–20		
10520+560		10 52 00 56 00 00	151.11 54.55	30.8000		R	0.2–0.4		−
10528+606	3U 1109+59 2A 1052+606	10 52 31 60 42 29	145.43 51.40	0.4630	12.70–3.71	U	2–6	4.0	−
					1974–76	A	2–18	1.3	−
					5–10.11.77	H	2–10	4.2	−
					2–9.5.1978	H	2–10	3.2	upper limit
10586−226	2A 1058−226 4U 1057−21	10 58 38 −22 40 12	272.26 33.28	0.3770	1971–73	U	2–6	1.45	−
					1974–76	A	2–18	2.3	−
10598+384		10 59 48 38 28 12	180.12 64.48		23–25.11.77	H	0.2–2.0		−
11015+450		11 01 33 45 02 00	165.79 62.12		16–26.11.75	S3	0.1–0.4		−
11016+384	2A 1102+389 A 1103+38	11 01 39.7 +38 28 51	179.92 65.11	0.0004	1974–76	A	2–18	3.7	−
					1974–76	A	2–18	85	burst
					25.4.1976	S3	2–6	9.09	−
					26.4.1976	S3	2–6	6.36	−
					27.4–1.5.78	S3	2–6	0.59	−
					5.1978	H	3–6	2.4	50
11095+597	1M 1109+59	11 09 31 59 42 00	143.89 53.49	6.3000	1971–73	O7	3–10	1.6	−
11102−580	4U 1110−58	11 10 12 −58 00 00	290.20 2.13	0.1410	1971–73	U	2–6	1.82	−
11110−603		11 11 03 −60 21 00	291.16 0.01	72.5000	1976	R	0.37–1.9		soft, burst
11189−615	A 1118−61	11 18 59 −61 35 18	292.51 −0.83	0.0065	1971–73	O7	3–10	18.0	upper limit
					17.12.74– 13.1.75	A	2.9–7.6	40–200	−
					18–31.1.75	A	2.9–7.6	20–40	transient
					23.1–3.2.76	A	2.9–7.6	100	−

Table I (continued)

Name XRS	Other names	R.A. decl. (1950)	l b	Area (sq. deg.)	Date	Range keV	Intensity μJy 2–6 keV	Var. comments
(1)	(2)	(3)	(4)	(5)	(6)	(7)	(8)	(9)	(10)
11190−603	4U 1118−60	11ʰ19ᵐ03ˢ	292.07	optical	1971−73	U	2–6	330	10
	Cen X-3	−60°20′54″	0.36	data	1971−73	O7	3–10	62.8	−
					11.1974	A	2–18	360	10
					7.1975	O8	2–6	270	var
11196−778	4U 1119−77	11 19 36	298.20	0.1967	1971−73	U	2–6	1.57	−
		−77 48 00	−16.04						
11206−431	4U 1120−43	11 20 36	286.37	0.6365	1971−73	U	2–6	1.23	−
		−43 10 30	16.56						
11222−590	H 1122−59	11 22 12	292.24	0.3910	11−20.1.78	H	1–10	4.7	−
		−59 03 00	1.16						
11304−146	4U 1130−14	11 30 24	276.18	2.2475	1971−73	U	2–6	2.02	−
		−14 37 30	43.85		1974−76	A	2–18	1.22	upper limit
11353+525		11 35 23	146.61	?	22−24.11.77	H	0.2–2.8		−
		52 33 06	61.34						
11357−373	2A 1135−373	11 35 46	287.26	0.1180	1971−73	U	2–6	3.29	−
	4U 1136−37	−37 21 36	23.00		1974−76	A	2–18	5.0	−
11373−651	3U 1134−61	11 37 18	295.54	0.0310	1971−73	U	2–6	16.0	−
	4U 1137−65	65 06 00	−3.54		1971−73	O7	3–10	6.7	upper limit
					10−25.11.74	A	2–18	2.3	upper limit
11432−185	A 1143−18	11 43 12	281.87	1.1000	1974−76	A	2–18	2.7	−
		−18 30 00	41.72						
11440+197	3U 1144+19	11 44 05	236.86	0.1300	1971−73	U	2–6	4.23	−
	2A 1141+199	19 43 12	73.28		1971−73	O7	3–10	2.6	−
	4U 1143+19				1974−76	A	2–18	2.7	−
					1975−77	O8	2–20		
11440+840	4U 1144+84	11 44 00	125.00	2.0182	1971−73	U	2–6	3.77	−
		84 00 00	33.15						
11448−748	1M 1144−74	11 44 48	298.75	0.15	1971−73	O7	3–10	2.8	−
		−74 49 48	−12.76						
11455−619	4U 1145−61	11 45 30.6	295.60	0.00087	13.5.1970	R	1.4–20	98	−
	1S 1145−619	−61 55 40	−0.20		1971−73	U	2–6	120	10
					1971−73	O7	3–10	17.4	−
					1975	S3	2–6	113	5
					1974−76	A	2–18	37	4
					7−20.5.78	S3	2–6	1060	
11474−124	4U 1147−12	11 47 26	280.45	2.1110	1971−73	U	2–6	1.04	−
		−12 24 36	47.48		1974−76	A	2–18	1.04	upper limit
11500+720	2A 1150+720	11 50 05	129.31	0.101	1974−76	A	2–18	3.7	−
		72 04 48	44.59		1975−77	O8	2–20		−
11508+745		11 50 48	128.21	50.24	14.5.72	V5	11–100	16700	burst
		74 30 00	42.31						
11536−115	4U 1153−11	11 53 36	282.14	0.2943	1971−73	U	2–6	1.8	−
		−11 33 00	48.80						
11539−402	4U 1153−40	11 53 54	291.75	0.2555	1971−73	U	2–6	2.59	−
		−40 12 00	21.20						
11576−071	A 1157−07	11 57 36	281.57	2.5	1974−76	A	2–18	2.48	−
		−07 06 00	52.88						
12036−061	4U 1203−06	12 03 41	283.22	0.3465	1971−73	U	2–6	2.0	−
		−06 07 12	54.73						
12070−521	H 1207−52	12 07 00	296.55	optical	17−21.1.78	H	0.25–1		−
		−52 07 00	9.52	data					
12078+397	3U 1207+39	12 07 53	154.91	0.0110	1971−73	U	2–6	5.5	−
	2A 1207+397	39 47 24	74.96		1971−73	O7	3–10	7.9	−
	4U 1206+39				1974−76	A	2–18	24.7	2
					25.5.77−5.6.77	O8	2–60	3.7	2
12096−452	4U 1209−45	12 09 36	295.76	0.2477	1971−73	U	2–6	1.52	−
		−45 12 00	16.85						

Table I (continued)

Name XRS	Other names	R.A. decl. (1950)	l b	Area (sq. deg.)	Date		Range keV	Intensity μJy 2–6 keV	Var. comments
(1)	(2)	(3)	(4)	(5)	(6)	(7)	(8)	(9)	(10)
2103 − 646	4U 1210 − 64	12ʰ10ᵐ22ˢ −64°38′24″	298°88 −2°35	0.019 5	·1971–73 1971–73 10–25.11.74	U O7 A	2–6 3–10 2–18	7.5 8.6 4.5	− − −
2150 + 440		12 15 00 44 00 00	143.18 72.04		1976	R	0.2–0.4		soft, burst
2156 − 594	A 1215 − 59	12 15 36 −59 24 00	298.75 2.92	0.0570	10–25.11.74	A	2–18	9.0	2
2170 − 672		12 17 00 −67 12 00	293. −7.	113	18.10.1969	V5	3–12	16700	burst
2190 + 305	2A 1219 + 305	12 19 00 +30 31 01	186.36 82.66	0.0280	1974–76 5.1978	A H	2–18 3–6	12.7 1.9	20 4
2200 + 260		12 20 00 26 00 00	223.56 83.37	73.8	1976	R	0.1–0.4		soft, burst
2214 − 081	3U 1237 − 07 4U 1221 − 08	12 21 24 −08 09 00	291.38 53.83	0.2427	1971–73 1974–76	U A	2–6 2–18	2.29 4.8	− upper limit
2238 − 624	3U 1223 − 62 GX 301 + 0 GX 301 − 2	12 23 50.2 −62 29 35	300.11 −0.10	0.00087	1971–73 1971–73 10–25.11.74 10.1975 1.1976 4.1976 23.1.78	U O7 A C O8 S3 S3	2–6 3–10 2–18 3–18 18–36 2–11 6–14	67 41.7 130 175 25 1000	5 3 var burst
2260 + 024	3U 1224 + 02 2A 1225 + 022 4U 1226 + 02	12 26 33.2 02 19 27	289.61 64.48	0.0262	1971–73 1971–73 1974–76 6–7.78	U O7 A H	2–6 3–10 2–18 1.5–13.5	3.19 3.2 6.3 4.5	3.5
2270 + 198		12 27 00 19 48 00	270. 81.	113	24.10.1969	V5	3–12	3840	burst
2287 + 126	Vir X-1 4U 1228 + 12 2A 1228 + 125	12 28 05 12 42 00	283.56 74.51	0.0211	1971–73 1971–73 1974–76 1975–77 12.1977	U O7 A O8 H	2–6 3–10 2–18 2–20 0.2–2.5	36.2 21.6 35	
2329 + 071	3U 1231 + 07 4U 1232 + 07	12 32 54 07 06 00	291.61 69.34	0.5272	1971–73 1971–73 1974–76	U O7 A	2–6 3–10 2–18	1.5 3.0 5.3	− − upper limit
2391 − 599	A 1238 − 59	12 39 07.5 −59 55 39	301.8 2.6	0.0002	19.12.74– 13.1.75 28.11–2.12.75 5–12.12.1975 23.1–3.2.1976 1975–77	A A A A S3	2.9–7.6 2.9–7.6 2.9–7.6 2.9–7.6 2–11	80 40 40 80 15	upper limit 2 4 −
2403 − 056	4U 1240 − 05	12 40 22 −05 39 00	299.06 56.88	6.2847	1971–73 1974–76	U A	2–6 2–18	1.0 2.2	− −
2446 − 603	A 1244 − 60	12 44 38 −60 22 12	302.46 2.23	0.0335	19.12.74– 2.1.75 2–3.1.75 3–13.1.75	A A A	2.9–7.6 2.9–7.6 2.9–7.6	60 160 50	upper limit bursts upper limit
2462 − 410	3U 1247 − 41 2A 1246 − 410 4U 1246 − 41	12 46 13 −41 02 06	302.44 21.56	0.0149	1971–73 1971–73 1974–76 1975–77	U O7 A O8	2–6 3–10 2–18 2–20	8.55 8.1 10.8	− − − −
2466 − 588	A 1246 − 58 4U 1246 − 58	12 46 39 −58 51 00	302.70 3.75	0.0176	1971–73 1971–73 10–25.11.74 1.1975 21.7–16.8.75 11.75–2.76	U O7 A A A A	2–6 2–6 2–18 2.9–7.6 2.9–7.6 2.9–7.6	3.67 167 13 480 13 40	− transient 3 upper limit

P. R. AMNUEL ET AL.

Table I (continued)

Name XRS	Other names	R.A. decl. (1950)	l b	Area (sq. deg.)	Date		Range keV	Intensity μJy 2–6 keV	Var. comments
(1)	(2)	(3)	(4)	(5)	(6)	(7)	(8)	(9)	(10)
12492 − 289	3U 1252 − 28	12ʰ49ᵐ41ˢ6	303°19	0.000 004	1971–73	U	2–6	6.26	
	4U 1249 − 28	−28°59′07″	34°91		1971–73	O7	3–10	5.8	
	2A 1251 − 290				1974–76	A	2–18	8.2	
					1975–77	O8	2–20		
					7–13.9.77	H	2–10	3.3	
					13–16.1.78	H	2–10	4.48	
					5–12.3.78	H	2–10	4.3	
					7.7.1978	H	2–10	12	
12500 − 667	A 1250 − 66	12 50 00 −66 45 00	303.10 −4.15	0.290 0	10–25.11.74	A	2–18	9.0	−
12539 − 002	4U 1253 − 00	12 53 54 −00 16 30	305.64 62.30	0.269 4	1971–73	U	2–6	1.25	−
12543 − 690	4U 1254 − 69	12 54 15.8 −69 01 20	303.47 −7.06	0.000 12	1971–73	U	2–6	38.7	−
					1971–73	O7	3–10	39.6	−
					11.1974	A	2–18	63	1.4
					1975–77	S3	2–11	43.4	−
					8.1977	H	2–6	4.5	−
12544 − 169	A 1254 − 16	12 54 24 −16 57 36	304.91 46.86	1.700 0	1974–76	A	2–18	2.43	−
12566 − 171	H 1256 − 171	12 56 41 −17 07 48	305.62 45.44	0.820 0	12.1.1978	H	2 − 10	1.37	−
12574 + 283	Coma X-1	12 57 24 28 18 00	59.24 87.93	0.005 0	1971–73	U	2–6	24.7	−
	3U 1257 + 28				1971–73	O7	3–10	18.1	−
	2A 1257 + 283				1974–76	A	2–18	30.1	−
	4U 1257 + 28				1975–77	O8	2–20		−
12582 − 613	4U 1258 − 61	12 58 14.3 −61 20 01	304.08 1.24	0.000 8	1971–73	U	2–6	92	10
	GX 304 − 1				1971–73	O7	3–10	50.2	
	2S 1258 − 613				1974–76	A	2–18	67	5
					1977	S3	1–12	19	var
13007 − 488	4U 1300 − 48	13 00 42 −48 52 30	304.98 13.68	0.297 9	1971–73	U	2–6	2.8	−
13020 − 775	4U 1302 − 77	13 02 00 −77 30 00	303.73 −14.92	6.049 3	1971–73	U	2–6	1.77	−
13061 − 012	2A 1306 − 012	13 06 07 −01 16 48	311.86 61.03	0.210 0	1974–76	A	2–18	5.0	−
13083 + 362		13 08 18.5 36 12 00	99.25 80.29	optical data	20.12.1977 19, 21, 12.77	H	0.2–2.8		soft
13088 + 864	4U 1308 + 86	13 08 48 86 28 48	122.65 30.91	0.142 5	1971–73	U	2–6	1.10	−
13108 + 371	H 1310 + 371	13 10 50 37 07 12	99.01 79.25	1.140 0	21.12.77	H	2–10	1.2	−
13140 + 292	MX 1313 + 29	13 14 00 29 15 00	84.20 83.60	0.150 0	15.6.1974	R	0.07–0.3		soft
					1975	S3	0.1		soft
13144 + 595	4U 1314 + 59	13 14 24 59 31 30	116.99 57.58	6.605 8	1971–73	U	2–6	5.93	−
13149 − 646	4U 1314 − 64	13 14 57 −64 36 00	305.78 −2.14	0.108 9	1971–73	U	2–6	16.08	−
13175 + 067	4U 1317 + 06	13 17 30 06 46 30	322.45 68.29	1.622 4	1971–73	U	2–6	1.2	−
13200 − 105		13 20 00 −10 31 00	315.37 51.70		1976	R	0.2–0.4		soft, burst
13223 − 427	4U 1322 − 42	13 22 22 −42 45 54	309.48 19.42	0.013 8	1971–73	U	2–6	12.7	−
	2A 1322 − 427				1974–76	A	2–18	78	4
					5–9.6.75	S3	2–6	27	var
					10.75, 5.76, 8.76, 2.77	A	2–10	5.9	−
13236 − 620	2A 1322 − 616	13 23 36 −62 00 54	307.05 0.31	0.027 8	1971–73	U	2–6	6.0	−
	4U 1323 − 62				10–25.11.74	A	2–18	13.36	3
	3U 1320 − 61								

Table I (continued)

Name XRS	Other names	R.A. decl. (1950)	l b	Area (sq. deg.)	Date		Range keV	Intensity μJy 2–6 keV	Var. comments
(1)	(2)	(3)	(4)	(5)	(6)	(7)	(8)	(9)	(10)
13240 − 625	Cen X-2	13^{h}24^{m}00^s −62°30′00″	307°1 0°3	16	28.10.1965	R	2–5	300	upper limit
					4.4.1967	R	2–5	13000	transient
					10.4.1967	R	2–5	19000	maximum
					20.4.1967	R	2–5	8800	−
					18.5.1967	R	2–5	3000	−
					28.9.1967	R	2–5	370	upper limit
					12.9.1968	R	2–5	120	upper limit
					3.11.1968	R	2–5	800	second flare
					7.11.1968	R	2–5	970	−
					1971–73	U	2–6	8	upper limit
13255 − 020	H 1325 − 020	13 25 34 −02 02 24	321.10 59.27	1.10	12.1.1978	H	2–10	1.49	−
13264 + 119	4U 1326 + 11	13 26 24 11 54 00	334.32 72.19	2.266 7	1971–73 1974–76	U A	2–6 2–18	1.65 1.71	− upper limit
13268 − 311	MX 1329 − 31 2A 1326 − 311 4U 1325 − 31	13 26 50 −31 09 36	312.42 30.75	0.314 0	1971–73 1971–74 1974–75 1974–76	U O7 O8 A	2–6 3–10 2–20 2–18	3.49 9.8 5.5	− var − −
13327 − 336	H 1332 − 336	13 32 46 −33 42 00	313.31 28.03	0.550 0	25.1.78	H	2–10	2.12	−
13358 + 402	A 1335 + 40	13 35 48 40 15 00	89.03 73.91	3.0	1974–76	A	2–18	1.49	−
13442 − 609	4U 1344 − 60 A 1343 − 60 2A 1343 − 602	13 43 53 −60 15 36	309.76 1.62	0.067 0	1971–73 10−25.11.74	U A	2–6 2–18	3.77 6.68	− 1.5
13448 − 325	MX 1347 − 32 2A 1344 − 325 4U 1345 − 32	13 44 48 −32 33 00	316.38 28.60	0.101 0	1971–73 1971–74 1974–76	U O7 A	2–6 3–10 2–18	9.23 10.8 12.7	− var −
13464 − 300	2A 1347 − 300	13 46 27.87 −30 03 40.6	317.34 30.39	optical data	1974–76	A	2–18	7.7	−
13468 + 266	3U 1349 + 24 2A 1346 + 266 4U 1348 + 25	13 46 48 26 40 48	33.11 77.12	0.078 0	1971–73 1974–76 1977–78	U A H	2–6 2–18 2–6	6.09 5.0 1.75	− − −
13481 + 700	2A 1348 + 700	13 48 10 70 01 48	115.72 46.53	0.275 0	1974–76	A·	2–18	4.0	−
13505 + 390	H 1350 + 390	13 50 31 39 02 24	79.16	1.020 0	29.12.77	H	2–10	1.24	var?
13522 + 187		13 52 12 18 42 00	5.47 73.05		17.6.76	R	0.15–0.28		−
13539 − 645	MX 1353 − 64	13 53 54 −64 30	310.5 −2.9	0.060 0	29.9.71− −18.5.74	O7	2–6	97.4	var
					12−25.11.74	A	2–18	5.0	−
					21.7−16.8.75	A	2–18	1.3	upper limit
					9−12.1.1976	A	2–18	4.0	upper limit
14010 − 452	MX 1401 − 45	14 01 00 −45 12 00	316.2 15.7	0.4	15.7.1975	O7	0.1–0.28		soft
14044 + 145	4U 1404 + 14	14 04 26 14 35 24	0.69 68.28	1.306 1	1971–73 1974–76	U A	2–6 2–18	1.54 0.75	− upper limit
14069 − 619	MX 1406 − 61	14 06 55 −61 54 00	312.04 −0.67	0.148	1971–73	O7	3–10	23.8	−
					12−25.11.74	A	2–18	2.7	upper limit
					21.7−16.8.75	A	2–18	1.3	upper limit
					9−12.1.1976	A	2–18	2.7	upper limit
14104 − 619		14 10 24 −61 54 33	312.10 −0.80	28.26	12.71−4.72	O7	10	22.4	transient
					1971–73	U	2–6	17 8	transient upper limit
14106 − 029	4U 1410 − 03 2A 1410 − 029	14 10 39.1 −02 58 15	339.17 53.70	0.000 97	1971–73 1974–76 12−15.5.77 20−21.5.77	O7 A S3	3–10 2–18 2–11	3.0 5.8 3.6	− − −

Table I (continued)

Name XRS	Other names	R.A. decl. (1950)	l b	Area (sq. deg.)	Date	Range keV	Intensity μJy 2–6 keV	Var. comments
(1)	(2)	(3)	(4)	(5)	(6)	(7)	(8)	(9)	(10)
14156+255	4U 1414+25 2A 1415+255	14^{h}15^{m}38^s 25°33′36″	32°52 70°55	0.114	1971–73 1974–76	U A	2–6 2–18	3.97 3.7	— —
14162−589	2A 1416−589	14 16 12 −58 58 12	314.04 1.75	0.132	1974–76	A	2–18	2.3	
14163−622	4U 1416−62	14 16 18 −62 16 30	312.96 −1.37	0.025 5	1971–73	U	2–6	12.67	—
14174−624	2S 1417−624	14 17 26.2 −62 28 14	313.12 −1.35	0.000 2	25–30.7.78	S3	2–11	50	3
14185−614	MX 1418−61 4U 1425−61	14 18 29 −61 21 00	313.51 −0.59	0.146	1971–73 1971–73 12–25.11.74 21.7–16.8.75 9–12.1.76	U O7 A A A	2–6 3–10 2–18 2–18 2–18	4 13.2 4.0 1.3 2.7	— upper limit upper limit
14186+485	2A 1418+485	14 18 41 48 32 24	89.65 62.66	0.223	1974–76	A	2–18	1.8	—
14260−624		14 26 00 −62 28 00	314.05 −1.70	optical data	21.7.75	Apollo	0.065–0.29		soft, flare
14362−606	H 1437−61	14 36 12 −60 38	315.78 −0.71	optical data	19–25.8.77	H	0.25		—
14365−566	4U 1436−56	14 36 34 −56 36 36	317.43 2.95	0.777 6	1971–73	U	2–6	1.02	—
14369−620	Cen XR-1	14 36 54 −62 00 00	315.4 −2.0	1	25.4.65 13.5.70	R R	1.5–8 0.2–2.7	140	— —
14384−185	4U 1438−18	14 38 24 −18 30 00	336.06 37.00	0.782 5	1971–73	U	2–6	1.08	—
14390−617	4U 1425−61 2A 1439−627 A 1439−61	14 39 02.4 −62 43 12	315.24 −2.75	0.176	10–25.11.74 21.7–16.8.75	A	2–18	4.5	—
14391−622		14 39 08 −62 15	315.38 −2.49	optical data	1971–73 7.1975	O7 B	3–10 0.6–2.5	7.9	SNR
14446+430	3U 1443+43 4U 1444+43	14 44 36 43 04 30	74.47 61.89	1.234 9	12.70–3.71 1971–73 1974–76	U O7 A	2–6 3–10 2–18	2.84 2.8 1.67	— upper limit upper limit
14484−556	4U 1446−554 2A 1448−554 LUp XR-1	14 48 24 −55 27	319.42 3.31	0.071	25.4.1965 17.5.1967 14.6.1969 13.5.1970 1971–73	R R R R U	1.5–8 2–16 1–10 1.4–20 2–6	160 960 60 50 3.52	transient? — — — —
14490+193		14 49 05 +19 18 27	23.09 61.35		22–24.1.78	H	0.2–2.8		—
14506−805	4U 1450−80	13 50 36 −80 33 00	308.05 −19.16	0.331 5	1971–73	U	2–6	2.10	—
14528−602	A 1452−60	14 52 48	317.8	0.30	12.70–7.72 10–25.11.74	U A	2–6 2–18	6.8 9.0	— 2
14550+191	4U 1455+19	14 55 02 +19 06 00	23.77 59.98	1.000 4	1974–76 1971–73	A U	2–18 2–6	1.30	—
14553−313	Cen X-4	14 55 19.5 −31 18 07	332.33 24.03	optical data	9.7.1969 11.7.1969 18.7.1969 20.9.1969 11–13.5.79 14.5.1979 17.5.1979 31.5.1979	V5 V5 V5 V5 A A A Corsa-B	3–12 3–12 3–12 3–12 3–6 3–6 3–6 1–12	10700 23400 9700 80 100 1490 4240 42400	transient maximum upper limit transient burst
14554−273	4U 1455−27	14 55 24 −27 22 12	334.60 27.38	7.250 3	1971–73	U	2–6	6.31	—
14566+225	4U 1456+22	14 56 41 22 35 24	30.70 60.75	1.428 1	1971–73 1974–76	U A	2–6 2–18	1.67 0.08	— —

Table 1 (continued)

Name XRS	Other names	R.A. decl. (1950)	l b	Area (sq. deg.)	Date		Range keV	Intensity μJy 2–6 keV	Var. comments
(1)	(2)	(3)	(4)	(5)	(6)	(7)	(8)	(9)	(10)
14580−415	4U 1458−41	14ʰ58ᵐ00ˢ −41°30′00″	327°43 14°92	0.241 7	1971–73 1971–73 1974–76 24–28.5.76	U O7 A S3	2–6 3–10 2–18 1–10	4.09 1.2 1.8 3.2	SNR upper limit
15059+573	4U 1505+57	15 05 54 57 18 00	93.70 51.61	0.328 4	1971–73	U	2–6	1.67	–
15087+062	2A 1508+062 MX 1514+06(?)	15 08 46 06 13 12	6.91 50.65	0.094	1971–73 1974–76 1975–77 1977–78	U A O8 H	2–6 2–18 2–20 2–6	7.01 6.3 2.4	– – – –
15100−390		15 10 00 −39 00 00	330.80 15.89	12.56	13.5.70 25.5.71 24–28.5.76	R R S3	0.2–2.7 0.6–1.6 0.15–0.5 0.5–0.8		SNR –
15101−590	4U 1510−59	15 10 07 −59 00 00	320.31 −1.21	0.0140	1971–73 1971–73 10–25.11.74	U O7 A	2–6 3–10 2–18	10.5 22.7 15.9	SNR? – –
15134+070	H 1513+070	15 13 26.4 07 05 24	9.09 50.77	1.62	3.2.1978	H	2–10	1.51	–
15142+068	MX 1514+06	15 14 12 06 48 00	8.5 50.2	0.786	1971–73 1971–73	O7 U	3–10 2–6	7.8	– –
15150+231	4U 1515+23	15 15 00 23 06 00	33.84 56.83	1.486 9	1971–73	U	2–6	1.15	–
15168−569	Cir X-1 Nor XR-2? 4U 1516−56 2S 1516−569	15 16 48.6 −50 59 14	322.11 0.05	0.000 097	25.4.65 14.6.1969 13.5.1970 5.69–8.70 1971–73 1971–73 8.1973 4.1974 11.1974 5.07.75 1.1976 17.2.76 18.2.76 1975–77 15.1.79	R R R V5 U O7 C C A Sal A A A S3 A	1.5–8 1–10 1.4–20 3–12 2–6 3–10 2.5–7.5 2.5–7.5 2–18 2–10 2–18 2–18 2–18 2–11 2–18	350 750 735 785 1200 75.3 38 350 9.5 5.9 2.7 184 9.0 1260 2120	– – – bursts 10 – 4 – – – upper limit burst upper limit 35 burst
15170−500		15 17 00 −50 00 00	326. 6.	113	26.10.1969	V5	3–12	3500	burst
15190−314		15 19 00 −31 24 00	337 21.	113	7.7.1969	V5	3–12	37200	burst
15190+082	2A 1519+082	15 19 00 08 12 00	11.71 49.71	0.282 0	1974–76	A	2–18	5.8	–
15212+285	4U 1521+28 2A 1518+274	15 21 17 28 31 12	44.03 56.50	1.928 6	1971–73 1974–76	U A	2–6 2–18	1.79 2.3	– –
15241−617	A 1524−61 TrA X-1 2S 1524−617	15 24 06.9 −61 42 41	320.30 −4.40	0.000 097	1971–73 12.11.1974 6–8.2.1975 6.1975	O7 A A S3	2–6 2–10 3–6 2–11	18.0 1500 170 81.8	upper limit transient
15307−443	4U 1530−44	15 30 46 −44 23 24	331.01 9.26	0.680 9	1971–73	U	2–6	4.96	–
15358−292	4U 1535−29	15 35 53 −29 13 12	341.39 20.71	1.262 0	1971–73	U	2–6	330	burst
15382−521	Nor 2? GX 327+4.5 4U 1538−52 2S 1538−522	15 38 14 −52 10 48	327.40 2.24	0.009 6	25.4.1965 18.5.67 16.6.1969 1971–73 1971–73	R R R U O7	1.5–8 2–16 1–10 2–6 3–10	220 1840 670 30 18	– – – 2 –

Table I (continued)

Name XRS	Other names	R.A. decl. (1950)	l b	Area (sq. deg.)	Date		Range keV	Intensity μJy 2–6 keV	Var. comments
(1)	(2)	(3)	(4)	(5)	(6)	(7)	(8)	(9)	(10)
					10–25.11.74	A	2–18	27	2
					1975–77	S3	2–11	13	–
					1974–77	O8	2–18		
					20–28.8.77	H	2–6	5.4	–
15400 – 320	H 1540 – 32	15ʰ40ᵐ00ˢ −32°00′00″	340°.18°.	?	20–28.8.77	H	0.15–1.8		SNR?
15412 – 534	A 1540 – 53	15 41 12 − 53 24 00	331.08 0.99	0.126 8	6–9.9.1976	A	1.8–5.6	6.7	var
					1977	O8		16.7	var
15430 – 624	4U 1543 – 62 2S 1543 – 624	15 43 33.9 − 62 24 56	321.71 − 6.29	0.000 2	1971–73	U	2–6	31.9	
					1971–73	O7	3–10	21.6	
					10–25.11.74	A	2–18	67	1.5
					1975–77	S3	2–11	57	
15438 – 475	4U 1543 – 47	15 43 50 − 47 33 36	330.93 5.36	0.000 6	26.7.1971	V5	3–12	7180	transient
					17.8.1971	U	2–6	3340	
					5.11.1971	O7	1–6	2170	
					10.12.1971	O7	1–6	710	
					23–25.11.71	O7	1–6	830	
					24–25.4.72	O7	1–6	130	
					12–25.11.74	A	2–18	21.2	
					9–12.1.1976	A	2–18	9.5	upper limit
15440 – 757	1M 1544 – 75	15 44 00 − 75 45 00	313.24 16.75	0.075 3	1971–73	O7	3–10	2.6	–
15452 – 536	MSH 15–56 = = GX 326.3 – 1.8	15 45 14 − 53 39 54	327.75 0.93	radio data		H	0.7–2		SNR
15539 – 542	MX 1553 – 54	15 53 55.6 − 54 16 15	328.0 − 0.9	0.000 29	15.6.1975	S3	2–11	47	transient
15561 – 756	2A 1556 – 756	15 56 06 − 75 39 36	313.89 − 17.17	0.662	1974–76	A	2–18	4.0	–
15564 – 527	A 1556 – 52	15 56 24 − 52 42 00	329.24 0.11	0.160 0	10–25.11.74	A	2–18	18	2
15565 + 272	3U 1555 + 27 2A 1556 + 274 4U 1556 + 27	15 56 34 27 14 06	44.04 48.60	0.061 6	1971–73	U	2–6	6.04	
					1971–73	O7	3–10	2.1	
					1974–76	A	2–18	6.3	
15569 – 606	4U 1556 – 60	15 56 47.2 − 60 35 47	324.13 − 5.97	0.000 2	10.7.1970	R	2.4–10.5	1000	
					1971–73	O7	3–10	23	
					1971–73	U	2–6	33	2
					1974–76	A	2–18	32	
16011 + 159	3U 1551 + 15 2A 1600 + 164 4U 1601 + 15	16 01 06 15 58 30	28.87 44.23	0.217 6	1971–73	U	2–6	3.07	–
					1974–76	A	2–18	3.7	–
					1975–77	O8	2–20		–
16040 – 590		16 04 00 − 59 00 00	326. − 5.	28.26	5.69–8.70	V5	3–12	620	bursts
16088 – 522	MX 1608 – 52 GX 331 – 1 4U 1608 – 52	16 08 53 − 52 17 29	330.95 − 0.81	0.000 08	6.69–8.70	V5	3–12	10690	bursts
					13.5.1970	R	1.4–20	42	–
					21.12.1971	U	2–6	1950	bursts
					22.12.1971	U	2–6	1320	burst
					11.5.1972	U	2–6	520	bursts
					1971–73	U	2–6	70	10
					1971–73	O7	3–10	100	2
					11.1974	A	2–18	2.3	upper limit
					11.1975	A	2–18	670	transient
					7.1977	S3	2–6	1900	transient
					30.8–8.9.77	H	1–15	3180	
					27.1.1979	A	3–6	53	upper limit
					31.1.1979	A	3–6	100	–
					2.2.1979	A	3–6	320	–
16128 + 339	σCrB	16 12 48.3 33 59 02	54.67 46.13		7–12.2.78	H	0.2–2.8		–

Table I (continued)

Name XRS	Other names	R.A. decl. (1950)	l b	Area (sq. deg.)	Date		Range keV	Intensity μJy 2–6 keV	Var. comments
(1)	(2)	(3)	(4)	(5)	(6)	(7)	(8)	(9)	(10)
16139 − 509	H 1615 − 51	16ʰ13ᵐ54ˢ −50°56′00″	333°26 0°46		2.3.1978	H	2–10	4.36	SNR
16140 − 395	GX 340 + 6 Sco XR-4	16 14 00 −39 31 00	340.40 7.3	0.502 4	25.4.1965 30.9.1965 7.7.1967 26.7.1968 14.6.1969	R R R R R	1.5–8 1–10 4–8 1.5–6 1.5–8	690 1300 750 1200 370	− − − − −
16146 − 277	4U 1614 − 27	16 14 36 −27 47 24	348.99 15.99	0.435 0	1971–73	U	2–6	2.29	−
16170 − 155	4U 1617 − 15 Sco X-1	16 17 04.3 −15 31 13	359.09 23.77	optical data	1971–73 1971–73 10.72–6.75	U O7 C	2.4–6.9 3–10 2.5–7.5	28390 13640 15030	5 − −
16186 + 150	North Polar Spur	16 18 36 15 04 58	30.00 40.00	great region	10.4.1973 27.8.1973	R R	0.6–2 0.1–1.6		SNR
16190 − 468		16 19 00 −46 48 00	336. 2.	113	27.10.1969	V5	3–12	10500	burst
16210 − 528	A 1621 − 52	16 21 00 −52 51 00	331.87 −2.53	0.170 0	10–25.11.74	A	2–18	5.8	−
16212 − 234	4U 1621 − 23	16 21 12 −23 27 00	353.37 17.85	0.452 0	1971–73	U	2–6	2.0	−
16220 − 243		16 22 00 −24 20 00	353.80 17.10	64		S3	0.1–0.8		−
16240 + 659	H 1623 + 659	16 24 00 65 57 00	97.85 39.36	4.5	15.12.1977	H	2–10	0.8	−
16243 − 490	4U 1624 − 49 2S 1624 − 490	16 24 19.6 −49 05 07	334.92 −0.27	0.000 3	25.4.1965 30.9.1965 13.5.1970 1971–73 1971–73 12–25.11.74 21.7–16.8.75 9–12.1.1976 1975–77	R R R U O7 A A A S3	1.5–8 4–8 1.4–20 2–6 3–10 2–18 2–18 2–18 2–11	690 920 230 84 70.2 36.1 5.3 5.3 73	transient? − − 5 − − upper limit upper limit −
16256 − 333	4U 1625 − 33	16 25 36 −33 18 00	346.55 10.48	2.485 1	1971–73	U	2–6	3.31	−
16270 − 092	4U 1627 − 09	16 27 00 −09 13 30	6.19 25.79	0.296 4	1971–73	U	2–6	3.16	−
16272 − 673	4U 1626 − 67 2A 1627 − 673 2S 1627 − 673	16 27 43.6 −67 21 23	321.75 −13.06	0.000 2	1971–73 1971–73 1974–76 4.1977	U O7 A S3	2–6 3–10 2–18 2–11	30 33.6 36.9 33	2
16278 + 396	3U 1639 + 40 2A 1626 + 396 4U 1627 + 39	16 27 48 39 36 00	62.86 43.52	0.099 2	1971–73 1974–76 1975–77 1977–78	U A O8 H	2–6 2–18 2–20 2–6	4.94 5.3 2.27	
16284 + 286	4U 1628 + 28 3U 1645 + 21	16 28 26 28 40 12	48.10 42.00	2.627 6	12.70–3.71 1971–73 1974–76	U U A	2.4–6.9 2–6 2–18	3.7 1.18	upper limit −
16301 − 472	4U 1630 − 47 Nor 1 (?) Nor XR-1 GX 337 + 0	16 30 11 −47 16 23	336.90 0.28	0.001 0	25.4.1965 14.6.1969 13.5.1970 12.70–3.71 10–11.1972 3.5.1974 1.1976 15.9.1977 17.9.1977 21.9.1977	R R R U U U A A A A	1.5–8 1–10 1.4–20 2–6 2–6 2–6 2–18 3–6 3–6 3–6	650 635 50 300 160 460 585 900 1060 1480	transient transient upper limit transient transient transient transient

Table I (continued)

Name XRS	Other names	R.A. decl. (1950)	l b	Area (sq. deg.)	Date		Range keV	Intensity μJy 2–6 keV	Var. comments
(1)	(2)	(3)	(4)	(5)	(6)	(7)	(8)	(9)	(10)
					11.1977	A	3–6	2170	
					2–7.3.1978	H	1–10	23	
					27.5.1978	A	2–18	530	
					6.6.1978	A	2–18	630	
					8.10.1978	A	3–6	350	
16315−643	3U 1632−64	16ʰ31ᵐ30ˢ	324°39	0.049 4	1971–73	U	2–6	7.65	−
	2A 1631−644	−64°19′48″	−11°40		1971–73	O7	3–10	11.3	−
	4U 1631−64				1974–76	A	2–18	14.9	−
16364+052	3U 1623+05	16 36 24	21.47	1.833 5	1971–73	O7	3–10	2.5	var?
	2A 1630+057	05 12 00	31.68		1971–73	U	2–6	2.0	−
	4U 1636+08				1974–76	A	2–18	1.3	−
16369−536	4U 1636−53	16 36 57.6	332.92	0.000 097	25.4.1965	R	2–8	680	
	MXB 1636−53	−53 39 21	−4.82		26.5.1971	R	2–18	350	
	2S 1636−536				1971–73	U	2–6	410	
	Nor X-1				1971–73	U	2–6	310	
					4.4.73, 15.4.1973	C	2–7	560	
					10–25.11.74	A	2–18	450	1.6
					1–4.1.1976	S3	2–6	3170	burst
					2–9.8.1976	O8	2–20	3170	burst
					7.1977	S3		540	
16400+400	MX 1640+40	16 40 00	64.00	?	1.7.1975	S3	0.1–0.28		soft
		40 00 00	41.00						
16408−463		16 40 48	338.93	0.24	13.5.1970	R	1.4–20	147	−
		−46 18 00	−0.41						
16410−326	H 1641−325	16 41 04.8	349.27	0.82	6.9.1977	H	2–10	0.8	−
		−32 36 00	8.52						
16420−343		16 42 00	350.00	28.26	5.69–8.70	V5	3–12	610	bursts
		−34 18 00	5.00						
16421−455	4U 1642−45	16 42 06	339.58	0.000 8	25.4.1965	R	1.5–8	1620	
	GX 340+0	−45 31 30	−0.08		30.9.1965	R	4–8	1500	
	Ara X-1				7.7.1967	R	1.5–6	1400	
	GX 340−2				26.7.1968	R	1.5–8	1260	
	2S 1642−455				14.6.1969	R	1–10	600	
					13.5.1970	R	1.4–20	610	
					1971–73	O7	3–10	550	
					1971–73	U	2–6	750	3
					10–25.11.74	A	2–18	850	1.3
					1975–77	S3	2–11	500	
16446+699	4U 1644+69	16 44 36	101.66	0.218 3	1971–73	U	2–6	1.32	−
		69 55 30	36.13						
16450−284	H 1645−284	16 45 00	353.03	0.39	7.9.1977	H	2–10	1.95	−
		−28 28 12	10.52						
16450−507		16 45 00	336.00	113	1.12.1969	V5	3–12	3170	burst
		−50 42 00	−4.00						
16450−575		16 45 00	330.00	28.26	5.69–8.70	V5	3–12	620	bursts
		−57 30 00	−8.00						
16460+284		16 46 00	49.00	113	1.10.1969	V5	3–12	7350	burst
		28 24 00	38.00						
16489−185	H 1648−185	16 48 58	1.59	1.85	6.9.1977	H	2–10	1.30	−
		−18 34 48	15.92						
16494−595	H 1649−595	16 49 29	329.5	0.84	12.9.1977	H	2–10	2.40	−
		−59 31 12	−9.92						
16518−065	4U 1651−06	16 51 48	12.46	0.457 5	1971–73	U	2–6	2.20	−
		−06 31 30	22.29						
16522+398	4U 1651+39	16 52 13	63.70	0.000 89	1971–73	U	2–6	2.71	−
		39 51 31	38.97		8.1977				
16528+635	4U 1652+63	16 52 48	93.75	0.111 6	1971–73	U	2–6	1.23	−
		63 33 00	37.16						

Table I (continued)

Name XRS	Other names	R.A. decl. (1950)	l b	Area (sq. deg.)	Date		Range keV	Intensity μJy 2–6 keV	Var. comments
(1)	(2)	(3)	(4)	(5)	(6)	(7)	(8)	(9)	(10)
16531 − 407	OAO 1653 − 40 V 861 Sco	16^{h}53^{m}06^{s}9 −40°44′44″	344°5 1°5		24–26.4.78	C	3–19	28	
16559 − 424		16 55 54 −42 26 00	343.54 0.02	0.04	30.9.1965 21.9.1968 13.5.1970	R R R	4–8 1.5–7 1.4–20	367 718 92	transient? — —
16560 + 354	Her X-1 4U 1656 + 35 3U 1653 + 35	16 56 02 35 25 03	58.26 38.12	optical data	1971–73 1971–73 1974–76 26.2.1975 26.1.1975 28.8.1975 7–8.1977	U O7 A ANS S4 O8 S3	2–6 3–10 2–18 0.2–0.28 2–10 2–20 1–12	167 28.0 294 164	10 — 65 soft — —
16589 − 298	MXB 1659 − 29	16 58 55.7 −29 52 02	354.69 7.91	0.000 114	2–9.10.1976 10–14.9.77 6–10.3.78 9.3.1978	S3 H H H	1–12 1.0–13.3 1–10 1–10	534 22.7 4.5	bursts — upper limit
16589 − 487	4U 1658 − 48 GX 339 − 4 MX 1658 − 48	16 59 02.2 −48 43 11	338.93 −4.32	0.000 1	1971–73 1971–73 10–25.11.74 21–23.6.78	O7 U A S3	1–6 2–6 2–18 1–6	534 584 334 424	60 3 5
16592 + 337	2A 1659 + 337	16 59 17 33 43 48	56.22 36.56	0.763	1974–76	A	2–18	2.7	—
16594 − 765	4U 1659 − 76 3U 1544 − 75	16 59 24 −76 33 00	315.75 −20.64	0.245 3	1971–73	U	2–6	2.72	—
17005 − 377	4U 1700 − 37	17 00 32.7 −37 46 27	347.75 2.19	optical data	1971–73 1971–73 6.7.1975 27–31.03.77	U O7 S3 S3	2–6 3–10 2–11 1–10	160 77.4 100	10 — —
17023 − 363	4U 1702 − 36 GX 349 + 2 GX-10.7 Sco XR-2	17 02 21 −36 21 54	349.09 2.75	0.000 3	16.6.1964 25.4.1965 30.9.1965 11.10.1966 1.10.1967 26.7.1968 3.11.1968 5.12.1968 14.6.1969 13.5.1970 1971–73 1971–73 7.1974 6–7.1975 7.1977	R R R R R R R R R R U O7 C S3 S3	1.5–8 1.5–8 4–8 4–8 1.5–8 2–10 1–10 1–10 1.4–20 2–6 3–10 2.5–75 2–11 2–11	1200 2230 4100 1400 2470 1030 2450 460 910 560 1250 650 1360 860 860	— — — — — — — — — — 2 — var — —
17023 − 429	GX-14.1 4U 1702 − 42 2S 1702 − 429	17 02 19 −42 58 48	343.84 −1.27	0.016 0	29.2.1968 13.5.1970 1971–73 1971–73 12–25.11.74 21.7–16.8.75 9–12.1.1976	B R O7 U A A A	17–42 1.4–20 3–10 2–6 2.4–19.8 2.4–19.8 2.4–19.8	147 93.6 50 4.0 14.9 12.2	— — — 3 upper limit — upper limit
17036 + 261	4U 1703 + 26	17 03 36 26 06 00	47.44 33.8	1.904 2	1971–73	U	2–6	1.99	—
17040 + 241	2A 1704 + 241 4U 1700 + 24	17 04 05 29 09 00	45.24 33.12	0.137 0	1971–73 1974–76	U A	2–6 2–18	7.0 7.7	2 var
17043 − 304	4U 1704 − 30	17 04 18 −30 24 00	354.11 6.02	0.763 5	1971–73	U	2–6	5.24	—
17045 − 320		17 04 31 −32 06 36	352.76 4.96	0.057 7	1971–73	O7	3–10	6.5	—
17051 − 431		17 05 06 −43 10 12	344.0 −1.8	several degrees	9.75–9.76	O8	2–20	1670	bursts

Table I (continued)

Name CRS	Other names	R.A. decl. (1950)	*l* *b*	Area (sq. deg.)	Date		Range keV	Intensity μJy 2–6 keV	Var. comments
1)	(2)	(3)	(4)	(5)	(6)	(7)	(8)	(9)	(10)
7051−250	H 1705−25	17^{h}05^{m}10^{s}2 −25°01′20″6	358°60 9°10	0.000 15	1971–73 7–10.8.77	U A	2–6 2–18	2.5 5010	upper limit transient
7052+609	2A 1705+609	17 05 12 60 54 00	90.17 36.23	1.620	1974–76	A	2–18	1.8	−
7054−440	4U 1705−44	17 05 24 −44 03 00	343.32 −2.36	0.000 8	29.2.1968 1971–73 1971–73	B O7 U	18–50 3–10 2–6	48.8 470	− − 3
7056−322	3U 1704−32 4U 1705−32	17 05 41 −32 13 12	352.82 4.70	0.025 7	1971–73	U	2–6	42	5
7064+321		17 06 24 32 06 00	54.64 34.76	9.8	1971–73	O7	3–10	2.5	−
7065−273		17 06 32 −27 18 00	357.00 7.40	0.400 0	27.8–1.9.76	A	2–10	67	−
7068−434		17 06 48 −43 24 00	344.11 −2.11	0.3	9.75–9.76 23–28.10.77	O8 S3	2–20 2–11	2000	bursts bursts
7074+786	3U 1706+78 2A 1705+786 4U 1707+78	17 07 24 78 40 30	110.96 31.74	0.021 9	1971–73 1971–73 1974–76 1975–77	U O7 A O8	2–6 3–10 2–18 2–20	39.7 4.2 5.3	− − − −
7083−407	MX 1709−40 4U 1708−40	17 08 22 −40 46 12	346.28 −0.84	0.007 2	1971–73 1971–73	U O7	2–6 3–10	24.9 46.1	− −
17089−232	3U 1709−23 Oph 2 (?) 4U 1708−23	17 08 59 −23 17 42	0.53 9.36	0.005 5	25.4.1965 1971–73 1971–73	R O7 U	1.5–8 3–10 2–6	430 26.6 50	− − 3
17100−510		17 10 00 −51 00 00	338.80 −7.00	28.26	5.69–8.70	V5	3–12	610	bursts
17104−303		17 10 24 −30 20 00	356.40 2.30	⁻0.600 0	11.9.1975 14–15.9.75	O8 O8	1.7–60 1.7–60	2330 33	burst −
17106−474	MX-12	17 10 36 −47 24 00	321.80 −5.10	37.62	21–25.6.71	Cos	30		hard
17108−340	A 1710−34	17 10 52 −34 00 36	352.03 2.76	0.006 8	14.2–1.3.75 25.2–28.3.1976 15–18.4.76 17–24.9.76	A A A A	2.9–7.6 2.9–7.6 2.9–7.6 2.9–7.6	40 70 70 200	upper limit − − −
17151−393	3U 1714−39 4U 1715−39 GX-10.7	17 15 07 −39 19 12	348.21 −1.03	0.058 6	1971–73 1971–73	O7 U	3–10 2–6	26.0 21.4	− −
17154+028	4U 1715+02	17 15 24 02 48 00	24.37 22.00	0.149 5	1971–73	U	2–6	1.3	−
17160−487		17 16 00 −48 42 00	340.60 −6.57	10	21.9.1968	R	1.5–6.6	2500	−
17161−318	MX 1716−31	17 16 06 −31 49 12	354.45 3.14	0.219 5	8.11.1971 4.1.1972 1971–73 5.1977	O7 O7 O7 S3	3–10 3–10 3–10 2–11	270 1000 45 40	burst burst 3 −
17166−016	4U 1716−01	17 16 38 −01 39 36	20.37 19.55	10.336 7	1971–73	U	2–6	3.0	−
17194+661		17 19 24 66 06 00	329.83 33.90	8.000 0	1974–76	A A	2–18 2–18	1.8 2.6	upper limit −
17196−347	Sco X-6	17 19 36 −34 42 00	352.60 0.90	1	30.9.1965 5.12.1968 13.5.1970	R R R	4–8 1–10 1.4–20	800 60 40	transient? − −
17202+346	4U 1720+34	17 20 12 34 36 00	58.35 32.51	2.881 0	1971–73 1974–76	U A	2–6 2–18	0.5 3.38	− upper limit
17228−305	4U 1722−30	17 22 50 −30 31 30	356.33 2.70	0.011 5	1971–73	U	2–6	7.48	burst

Table I (continued)

Name XRS	Other names	R.A. decl. (1950)	l b	Area (sq. deg.)	Date	Range keV	Intensity μJy 2–6 keV	Var. comments
(1)	(2)	(3)	(4)	(5)	(6)	(7)	(8)	(9)	(10)
17228 + 119	4U 1722 + 11	17^{h}22^{m}48^s 11°57′00″	34°16 24°48	0.0576	1971–73	U	2–6	2.79	–
17250 – 302	GX-2.5 GX 357+2.5	17 25 00 − 30 16 00	357.30 1.40	1.000	14.6.1969 13.5.1970	R R	1–10 1.4–20	139 35	transient? –
17270 – 335		17 27 22 − 33 42 00	354.24 0.13	0.0078	1971–73 10.3.1976	O7 A	3–10 2–7	148.4 310	bursts
17271 + 503		17 27 06 50 21 00	71.18 33.54	optical data		H	3–6	0.6	–
17276 – 214		17 27 41 − 21 27 00	4.52 6.82	radio data	11–17.3.76	O8	0.5–1.5		–
17286 – 337	GX-5.6 GX 354+0 3U 1727−33 MXB 1728−34 2S 1728−337 4U 1728−33	17 28 39.6 − 33 47 52	354.31 −0.15	0.0002	1971–73 3–4.1976 1975–77 3.1978	U S3 S3 H S3	2–6 2–11 2–6 2–150	250 2500 200 15620	5 bursts var –
17288 – 162	4U 1728−16 GX 9+9 Oph 3	17 28 48.5 − 16 15 20	8.49 9.03	0.00004	1971–73 7.7.1967 26.7.1968 12.10.1969 1971–73 1971–73 15.4.1973 21.3.1975 1976 1975–77	OSO-7 R R R U O7 C C Ari S3	3–10 1.5–6 1.5–8 2.5–8 2–6 3–10 2–7 2.5–7.5 2.5–10 2–11	280 230 950 360 430 280 380 360 430 340	– 1.5 var var
17289 – 247	4U 1728−24 GX 1+4 GX 2+5 2S 1728−247	17 28 58 − 24 42 52	1.91 4.82	0.00027	1971–73 9.1972 1971–73 1975–77	O7 C U S3	3–10 2–7 2–6 2–11	116.5 110 100 82	– 1.5 2 –
17300 – 370		17 30 00 − 37 04 00	351.28 −2.48	72.5		R	0.37–1.9		soft
17302 – 333	MXB 1730−335	17 30 12 − 33 22 30	354.8 −0.2	0.00196	3.4.1976 14–15.3.76 11.1976 3.1978	S3 ANS A H	1–18 2–18 2–60	1670 400 180 13600	bursts bursts steady? bursts
17300 – 238		17 30 00 − 23 48 00	3.00 5.00	28.26	5.69–8.70	V5	3–12	610	bursts
17309 – 220	4U 1730−22	17 30 56 − 22 00 07	4.47 5.89	0.0026	12.70–1973	U	2–6	200	10
17329 – 449		17 32 56 − 44 55 21	345.4 −6.9	1.5	6.1971	Cos	40–190		hard, bursts
17342 – 127	H 1734−127	17 34 14 − 12 45 00	12.81 10.13	4.80	17.9.1977	H	2–10	1.50	–
17353 – 444	GX 346−7 Sco XR-3 3U 1735−44 4U 1735−44 MXB 1735−44 2S 1735−444	17 35 19.5 − 44 25 22	346.07 −6.97	0.000097	16.6.1964 1971–73 5.7.1975 1–8.5.1977 5.1977 1978	R U Sal S3 S3 S3	0.8–12 2–6 2–10 2–11 2–11 3–12	351 387 3340 220 220	1.7 burst bursts
17354 – 284	4U 1735−28 GX 352+2	17 35 24 − 28 27 00	359.57 1.56	0.0396	1971–73 11.3.1971 12.4.1971 9–12.4.1971 11.1974 21.7–16.8.75 9–12.1.76	O7 U U U A A A	3–10 2–6 2–6 2–6 2–18 2–18 2–18	32.0 943 67 83 125 65 139	– transient upper limit upper limit – – –
17360 – 163		17 36 00 − 16 18 00	10.00 8.00	28.26	5.69–8.70	V5	3–12	620	bursts

Table I (continued)

Name XRS	Other names	R.A. decl. (1950)	*l* *b*	Area (sq. deg.)	Date		Range keV	Intensity μJy 2–6 keV	Var. comments
(1)	(2)	(3)	(4)	(5)	(6)	(7)	(8)	(9)	(10)
17400 − 463		17ʰ40ᵐ00ˢ −46°18′00″	345°00 −9°00	28.26	5.69–8.70	V5	3–12	620	bursts
17401 − 391		17 40 08 − 39 11 29	351.00 − 5.00	50	26.5.1971	B	0.37–1.9		−
17417 − 296	MXB 1742 − 297	17 41 42 − 29 40 00	359.28 − 0.27	0.015	10.2.1976	S3	2–11	3340	burst
17424 − 289	A 1742 − 28 MXB 1743 − 29?	17 42 26 − 28 59 56	359.93 − 0.04	0.001 76	1974 16.2.1975 1.3.1976 17.6.1976	A A A SL	2–10 2–18 2–18 2.2–10.2	2000 3000 60 9	transient upper limit
17426 − 292	MXB 1743 − 293	17 42 36 − 29 16 00	359.73 − 0.22	0.060 0	1971–73 1–2.1976	O7 S3	3–10 2–11	170 1670	− bursts
17428 − 294	A 1742 − 294 2S 1742 − 294	17 42 53.2 − 29 29 77	359.56 − 0.36	0.000 2	17.6.1976 9–10.1976 2–3.1976 4.1976 5–6.1977	SL 1501 A A A S3	2.0–10.0 2.9–7.6 2.9–7.6 2.9–7.6 2–11	50 260 300 140 90	− − 6 bursts −
17430 − 295	A 1743 − 29	17 43 01 − 29 31 00	359.60 − 0.42	0.022 0	22.3.1976	A	2–18	250	transient
17436 − 291	GCX Sgr X-1 4U 1743 − 29 A 1742 − 294	17 43 36 − 29 07 48	359.95 − 0.33	0.091 7	23.4.1968 13.5.1970 1971–73	B R U	31–544 4–20 2–6	220 70	− − −
17436 − 285	MXB 1743 − 28	17 43 36 − 28 30 00	0.50 0.00	0.280 0	1976	S3	2–11	1670	bursts
17437 − 316	H 1743 − 32	17 43 46 − 31 40 52	357.80 − 1.70	6.000 0	12.8.1977 19.8.1977 22.8.1977 31.8.1977 3.9.1977 5.9.1977 8.77, 4.78 12–18.3.78	A A A A A A H H	3–6 3–6 3–6 3–6 3–6 3–6 1 1–10	150 640 950 600 950 1030 100 45.4	upper limit transient − − − − − −
17438 − 288	GX +0.2 −0.2	17 43 52 − 28 52 16	0.20 − 0.20	0.001 66	17.6.1976 15–19.2.75 24.2–1.3.75 29.2–27.3.76 15–25.4.76 16.9–7.10.76 9–22.4.77 1–4.5.77	SL 1501 A	2.2–10.2 2.9–7.6	40	−
17441 − 284	MXB 1744 − 28	17 44 06 − 28 28 00	0.59 − 0.07	0.020 0	31.3.1976	S3	2–11	2000	bursts
17448 − 361	A 1744 − 36	17 44 52 − 36 06 30	354.13 − 4.19	0.000 87	1971–73 2–3.1976 4.1976 9–10.1976	U A A A	2–6 2.9–7.6 2.9–7.6 2.9–7.6	16.7 400 40 160	upper limit transient? − upper limit
17448 − 265	4U 1744 − 26 GX 3 + 1 GX + 2.6	17 44 52 − 26 33 42	2.27 0.80	0.000 034	14.6.1964 30.9.1965 11.10.1966 7.7.1967 26.7.1968 5.12.1968 14.6.1969 13.5.1970 1971–73 1971–73 5.7.1975 1975–77	R R R R R R R R O7 U Sal 4 S3	1.5–8 4–8 2–5 1.5–6 1.5–8 1–10 1–10 1.4–20 3–10 2–6 2–6 2–11	1360 1110 1050 360 340 350 230 530 450 1000 1700 600	− -- -- − − − − − − 3 − −

Table I (continued)

Name XRS	Other names	R.A. decl. (1950)	l b	Area (sq. deg.)	Date		Range keV	Intensity μJy 2–6 keV	Var. comments
(1)	(2)	(3)	(4)	(5)	(6)	(7)	(8)	(9)	(10)
17453 + 390	4U 1745 + 39	17^{h}45^{m}19^s	64°.65	1.320 9	1971–73	U	2–6	1.2	—
	3U 1736 + 43	39°00′00″	28°.61		1971–73	O7	3–10	4.0	—
					1974–76	A	2–18	6.3	upper limit
17456 + 292	4U 1745 + 29	17 45 36	54.06	1.667 3	1971–73	U	2–6	1.75	—
		29 12 00	25.86						
17460 − 203	4U 1743 − 19	17 46 02	7.72	3.104 1	12.71–1.72	U	2–6	250	transient
		− 20 22 12	3.76		1971–73	O7	3–10	53.9	—
17468 − 370	4U 1746 − 37	17 46 48.7	353.55	0.000 2	1971–73	O7	1–10	92	2
		− 37 02 17	− 4.99		2–5.8.1977	S3	1–10	50	bursts
					1971–73	U	2–6	67	1.5
					6.75, 6.77	S3	1–10	13	—
17477 − 293	GX + 0.2, − 1.2	17 47 45	+ 0.21	0.320 0	17.6.1976	SL 1501	2.2–10.2	65	—
		− 29 23 00	− 1.23						
17491 − 284	GX + 1.1, − 1.0	17 49 06	1.12	0.020 6	15–19.2.1975				
		− 28 29 41	− 1.03		24.2–1.3.75				
					29.2–27.3.76	A	2.9–7.6	7	upper limit
					15–24.4.1976				
					16.9–7.10.76				
					9–22.4.1977				
					1–4.5.1977				
					17.6.1976	SL 1501	2.2–10.2	59	—
17522 − 008	H 1752 − 008	17 52 17	25.57	1.820 0	21.9.1977	H	2–10	1.26	—
		− 00 52 48	12.15						
17536 + 151	z Her	17 53 36	40.87		21–23.9.77	H	0.2–2.8		—
		15 08 54	18.48		19–20.3.78				
17555 − 338	Sco X-6	17 55 34	357.24	0.014 4	30.9.1965	R	4–8	750	—
	4U 1755 − 33	− 33 48 00	− 4.91		5.12.1968	R	1–10	57	—
	GX − 2.5				13.5.1970	R	1.4–20	35	—
					1971–73	U	2–6	100	2
					1971–73	O7	3–10	64.8	—
					6.1975	S3	2–11	94	—
17585 − 205	3U 1758 − 20	17 58 34	9.07	0.000 056	1.10.1964	R	4–8	2980	—
	GX 9 + 1,	− 20 32 00	1.15		25.4.1965	R	1.5–8	2330	—
	GX + 9.1				30.9.1965	R	4–8	2170	—
	Sgr X-3				11.10.1966	R	2–5	530	—
					7.7.1967	R	1.5–6	560	—
					25.8.1967	R	1–10	660	—
					26.7.1968	R	1.5–8	1030	—
					5.12.1968	R	1–10	430	—
					14.6.1969	R	1–10	600	—
					25.5.1971	R	2–18	1150	—
					1971–73	O7	3–10	540	—
					1971–73	U	2–6	1000	3
					4.4.1973	C	2–7	1380	—
					1975–77	S3	2–11	710	—
17590 − 664	4U 1759 − 66	17 59 00	327.56	0.327 7	1971–73	U	2–6	3.26	—
		− 66 27 00	− 20.19						
17590 − 250	4U 1758 − 25	17 59 02	5.08	0.000 6	1.10.1964	R	2.5–16	1930	—
	GX 5 − 1	− 25 04 30	− 1.03		30.9.1965	R	4–8	2620	—
	GX + 5.2				11.10.1966	R	2–5	1930	—
					7.7.1967	R	1.5–6	830	—
					25.8.1967	R	1–10	1260	—
					26.7.1968	R	1.5–8	1880	—
					14.6.1969	R	1–10	760	—
					25.5.1971	R	1.4–20	1010	—
					1971–73	O7	3–10	680	—
					1971–73	U	2–6	1920	2
					4.1973	C	2–7	2420	var
					5.7.1975	Sal		2080	—

Table I (continued)

Name XRS (1)	Other names (2)	R.A. decl. (1950) (3)	l b (4)	Area (sq. deg.) (5)	Date (6)	(7)	Range keV (8)	Intensity μJy 2–6 keV (9)	Var. comments (10)
18004+682	H 1800+682	18ʰ00ᵐ29ˢ 68°12′36″	98°28 29°77	1.8000	22.9.1977	H	2–10	0.50	–
18010+698		18 01 00 69 49 48	100.15 29.71	4.5	22.9.1977	H	2–10	0.5	–
18025–448		18 02 30 –44.51 36	348.10 –11.40	8.0000	6.1971	Cos	40–190		hard, bursts
18034–605	4U 1803–60	18 03 26 –60 33 36	333.50 –18.34	6.6050	1971–73	U	2–6	2.3	–
18037–246	MX 1803–24	18 03 47 –24 36 18	6.3 –2.0		12.70–1973	U	2–6	3	upper limit
					13.5.1976	S3	2–11	1670	transient
					19.5.1976				
18055–186		18 05 30 –18 37 00	11.60 0.70	?	11.1976	A	2–18	32	–
18061+458		18 06 06 45 50 57	73.16 26.44	optical data	19–25.2.77	A	260–1200		hard
18063–274		18 06 18 –27 24 00	4.00 –4.00	several degrees	22–25.3.76	O8	2–20	1670	bursts
18070–365		18 07 00 –36 30 00	356.00 –8.00	113	11.2.1970	V5	3–12	4170	burst
18079–108	4U 1807–10	18 07 55 –10 52 48	18.60 3.93	5.2676	6.1973	U	2–6	17	transient
18117–171 GX 13+1	Sgr XR-2 –17 10 16 GX+13.5 4U 1811–17	18 11 37.2 0.08	13.52	0.0001	16.6.1964	R	1–10	830	–
					1.10.1964	R	4–8	3170	–
					25.11.1964	R	1.5–8	1280	–
					30.9.1965	R	4–8	1100	–
					11.10.1965	R	2–5	560	–
					7.7.1967	R	1.5–6	230	–
					25.8.1967	R	2–10	530	–
					26.7.1968	R	1.5–8	600	–
					5.12.1968	R	1–10	180	–
					14.6.1969	R	1–10	500	–
					2.10.1969	R	1.5–8	180	–
					25.5.1971	R	2–18	500	–
					1971–73	O7	3–10	350	–
					1971–73	U	2–6	660	3
					5.7.1975	Sal 4		190	–
					1975–77	S3	2–11	480	–
					1975–77	C	3–8		–
18118+379	4U 1811+37	18 11 48 37 54 00	65.00 23.32	2.2208	1971–73	U	2–6	0.83	–
18124–121	4U 1812–12	18 12 26 –12 07 48	18.03 2.36	0.0268	1971–73	O7	3–10	27.1	–
					1971–73	U	2–6	33	2
18131–140	Ser XR-2 GX 17+2 GX+16.2 3U 1813–14 2S 1813–140	18 13 11.2 –14 03 10	16.42 1.48	0.0001	1.10.1964	R	2.5–16	1250	–
					25.4.1965	R	1.5–8	1720	–
					30.9.1965	R	4–8	2170	–
					11.10.1966	R	2–5	180	–
					7.7.1967	R	1.5–6	510	–
					25.8.1967	R	2–10	2000	–
					26.7.1968	R	1.5–8	1550	–
					5.12.1968	R	1–10	660	–
					14.6.1969	R	1–10	600	–
					2.10.1969	R	1.5–8	630	–
					25.5.1971	R	2–18	980	–
					1971–73	O7	3–10	700	–
					1971–73	U	2–6	1580	4
					1.8.1975	C	3–8	1100	var
					1975–77	S3	2–11	730	–
18149+498	3U 1809+50 A 1815+49 2A 1815+500	18 14 59 49 50 55	77.00 25.83	optical data	1971–73	U	2–6	3.79	
					1971–73	O7	3–10	2.5	
					1974–76	A	2–18	9.5	

Table 1 (continued)

Name CRS (1)	Other names (2)	R.A. decl. (1950) (3)	l / b (4)	Area (sq. deg.) (5)	Date (6)	(7)	Range keV (8)	Intensity μJy 2–6 keV (9)	Var. comments (10)
	4U 1813+50				11–12.10.75	O8	2–10	5	
					19–22.5.76	S3	0.1–0.8		var
					20.9.1977	B	16–120		
					22.9–12.10.77	H	0.1–3		
8159−083	A 1815−08	18ʰ15ᵐ55ˢ −08°21′36″	21°76 3°41	0.078 8	21.7–16.8.75	A	2–18	12.2	−
8162−123	A 1816−12	18 16 12 −12 18 36	18.32 1.47	0.008 8	21.7–16.8.75	A	2–18	24.4	−
8175−059	A 1829−06 4U 1817−05	18 17 34 −05 54 00	24.13 4.21	1.357 4	1971–73 10–25.11.74	U A	2–6 2–18	1.45 22.5	− 2.5
8204−303	Sgr X-4 Sgr 4 MXB 1820−30 2S 1820−303 4U 1820−30	18 20 28.4 31 23 14	2.78 −7.91	0.000 1	24.4.1965 30.9.1965 1971–73 1971–73 17–19.5.75 6.1975 7.1975 5.7.1975 27–30.9.75 9–12.1976 23–24.3.77	R R U O7 S3 S3 S3 Sal ANS A A	1.5–8 4–8 2–6 1–10 7–11 2–6 2–6 2–10 1–10 2.4–19.8 2–18	510 2170 530 400 3170 260 210 480 ⋅ 3670 62.6 350	3 5 burst
8217−054		18 21 42 −05 26 10	31.00 −8.00	several degrees	11–14.4.76	O8	2–20	830	−
8223−371	Sco XR-6 Sgr X-1 4U 1822−37 2A 1822−371	18 22 22.92 −37 08 09	356.85 −11.25	0.000 14	25.4.1965 30.9.1965 11.10.1966 26.7.1968 5.12.1968 1971–73 1971–73 1974–76 25–29.9.77	R R R R R O7 U A H	1.5–8 4–8 2–5 1.5–8 1–10 3–10 2–6 2–18 2–6	250 560 610 500 200 28.5 42 45.9 10	− − − − − − 4 2 −
8228−000	4U 1823−00 A 1822+00 2S 1822−000 3U 1822−00	18 22 48.5 −00 02 24	29.95 5.80	0.000 1	1971–73 1971–73 12–25.11.74 1977	O7 U A S3	3–10 2–6 2–18 2–11	45.4 90 90 43	− 2 2
18248+644	H 1824+644	18 24 53 64 25 12	94.16 27.09	1.200 0	16.11.1977	H	2–10	0.88	−
18251+339	4U 1825+33	18 25 07 33 56 24	61.93 19.48	1.384 5	1971–73	U	2–6	6.90	−
18288−065	A 1828−06	18 28 50 −06 31 12	24.89 1.45	0.066 1	21.7–16.8.75	A	2–18	20.2	−
18301+345	4U 1830+34	18 30 07 34 30 00	62.87 18.72	0.508 9	1971–73	U	2–6	5.49	−
18310−109	A 1831−10	18 31 00 −10 59 24	21.19 −1.11	0.049 3	21.7–16.8.75	A	2–18	14.9	
18317−232	3U 1832−23 4U 1831−23	18 31 47 −23 12 18	10.40 −6.90	0.019 7	1971–73 1971–73	U O7	2–6 3–10	10.9 11.1	− −
18325−051	4U 1832−05	18 32 30 −05 09 00	26.53 1.28	0.119 2	1971–73 1971–73 12–25.11.74	U O7 A	2–6 3–10 2–18	5.73 20.8 9.0	− − −
18332+326	H 1832+32	18 33 12.11 32 39 15.1	61.31 17.44	0.270 0	6.10.1977	H	2–10	3.0	−
18334−077	H 1833−077	18 33 24 −07 44 24	24.33 −0.12	0.02	2.10.1977	H	2–10	29.7	−
18341−626	H 1834−626	18 34 10 −62 38 24	332.75 −22.48	1.10	29.9.1977	H	2–10	1.17	−
18347−653	H 1834−653	18 34 43.2 −65 21 00	329.95 −23.28	0.700 0	29.9.1977	H	2–10	1.36	−

Table I (continued)

Name XRS	Other names	R.A. decl. (1950)	l b	Area (sq. deg.)	Date		Range keV	Intensity μJy 2–6 keV	Var. comments
(1)	(2)	(3)	(4)	(5)	(6)	(7)	(8)	(9)	(10)
18351−078	Sct X-1	18ʰ35ᵐ11ˢ −07°48′00″	24°41 −0°7	0.5119	23.6.1973 8.9.1974	R C	1.5–18 3–10	45.7 38	transient −
18353+387		18 35 18 38 42 00	67.43 19.17	0.0067	27.7.1977	R	0.15–0.8		soft
18358−114	A 1829−10 4U 1835−11	18 35 48 −11 24 00	21.37 −2.34	0.9078	10–25.11.74 1971–73	A U	2–18 2–6	13.5 4.2	3 −
18360−227		18 36 00 −22 42 00	11.20 −7.60	101.25	27, 25.3.76	O8	2–20	630	bursts
18374+049	Ser XR-1 GX+36.3 4U 1837+04 MXB 1837+049 2S 1837+049	18 37 29.8 04 59 23	36.12 4.85	0.0001	25.11.1964 25.4.1965 30.9.1965 11.10.1966 5.12.1968 24.9.1970 25.5.1971 1971–73 1971–73 12–25.11.74 4–7.10.1975 7.75, 7.76, 8.1976 1975–77 3.4, 6, 7, 9.78	R R R R R R R U O7 A S3 S3 S3 S3	1.5–8 1.5–8 4–8 2–5 1–10 1.2–16 2–18 2–6 3–10 2–18 2–24 2–11 2–11 3–12	600 600 360 230 70 380 400 460 210 310 1670 3500 360 7700	− − − − − − − 2 1.5 burst bursts burst
18383+629	H 1838+629	18 38 18 62 55 12	92.79 25.32	3.0000	16.11.1977	H	2–10	0.52	−
18389+378	H 1839+37	18 38 58 +37 52 24	66.86 18.24		10.1977	H	0.15–0.43		−
18400+013	A 1840+01	18 40 05 01 18 00	33.20 2.40	0.3000	21.7–16.8.75	A	2–18	22.5	
18434+675		18 43 26 67 30 00	97.88	2.1000	1971–73	O7	3–10	0.5	−·
18456−024	A 1845−02 2S 1845−024 4U 1850−03	18 45 41.1 −02 28 37	30.43 −0.40	0.0003	1971–73 10–25.11.74 1977	U A S3	2–6 2–18 2–11	16.7 40.6 15.7	4 2
18476−053	A 1847−05	18 47 36 −05 18 00	28.10 −2.10	0.2000	10–25.11.74	A	2–18	22.5	1.5
18476+789	3U 1825+81 4U 1847+78	18 47 36 78 54 00	110.53 26.91	0.9211	1971–73	U	2–6	2.2	
18482−079		18 48 15 −07 57 70	26.00 −3.80	several degrees	8.10.1976	O8	2–20	2300	burst
18490−771	1M 1849−77	18 49 00 −77 06 00	317.47 −26.65	0.1500	1971–73	O7	3–10	2.8	−
18492−312	4U 1849−31 −31 12 00	18 49 12	4.75 −13.83	0.3725	1971–73	U	2–6	4.57	−
18503−087	A 1850−08 2S 1850−087 4U 1850−08	18 50 21.9 −08 45 54	25.29 −4.18	0.0002	1971–73 10–25.11.74 1977	U A S3	2–6 2–18 2–11	11.8 40.6 20.7	3
18508+007	A 1850+00	18 50 48 00 42 00	33.80 −0.10	0.3000	10–25.11.74	A	2–18	27.0	2
18526+370	4U 1852+37	18 52 38 37 00 00	67.07 15.41	2.4189	1971–73	U	2–6	0.95	−
18536−239	4U 1856−23	18 53 36 −23 57 00	11.93 −11.72	2.9092	1971–73	U	2–6	4.79	−
18536+012	G 34.6−0.5	18 53 35 01 17 00	34.67 −0.43		8–9.10.1974 4–6.4.1975	ANS	1.0–3.5	3.6	SNR
18543+683	3U 1843+67 2A 1854+683 3U 1904+67 4U 1859+69	18 54 22 68 21 00	99.02 24.84	0.6400	1971–73 1974–76	U A	2–6 2–18	3.47 2.7	− −

Table I (continued)

Name CRS	Other names	R.A. decl. (1950)	l b	Area (sq. deg.)	Date		Range keV	Intensity μJy 2–6 keV	Var. comments
1)	(2)	(3)	(4)	(5)	(6)	(7)	(8)	(9)	(10)
9010+430	H 1901+43	19ʰ01ᵐ00ˢ 73°50'00"	43°00 16°20	10	18–22.10.77	H	0.18–0.43		var
9017+031	4U 1901+03	19 01 42 03 06 00	37.21 −1.39	0.0872	29.12.1970	U	2–6	145	−
					12.3.1971	U	2–6	42	−
					17.3.1971	U	2–6	50	−
					22.3.1971	U	2–6	33	−
					1971–73	U	2–6	145	10
					12.11.74–1.1976	A	2–18	2.7	upper limit
9048+670	1M 1904+67	19 04 48 67 00 00	97.83 23.56	1.1000	1971–73	O7	3–10	13.0	−
9059+000	A 1905+00 3U 1908+00 2S 1905+000 4U 1857+01 MXB 1906+00	19 05 54.9 00 05 37	35.00 −3.71	0.0003	1971–73	U	2–6	6.8	−
					21.7–16.8.75	A	2.4–19.8	22.5	1.5
					22–28.8.76	S3	2–11	7500	bursts
					1975–77	S3	2–11	14.4	−
9078+095	3U 1906+09 4U 1907+09	19 07 48 09 31 48	43.61 0.25	0.0620	1971–73	U	2–6	33	5
					1971–73	O7	3–10	20.1	−
					12–25.11.74	A	2–18	27.1	2
9082+023	Aql MXB	19 08 12 02 18 00	35.00 −3.20	1.3000	22–28.8.76	S3	2–11		burst
9087+005	Aql XR-1 4U 1908+00 2S 1908+005	19 08 43.6 00 30 05	35.70 −4.15	0.0001	25.4.1965	R	1.5–8	110	
					3.11.1968	R	2–10	810	
					21.9.1970	R	2–10	210	
					24.9.1970	R	1.2–16	33	
					1971–73	U	2–6	330	20
					1971–73	O7	3–10	52.4	
					1973	C	2–7	16.7	1.2
					12–25.11.74	A	2–18	5.0	
					4.1975	A	2–18	112.7	
					6.1975	A	2–18	1670	
					5.7.1975	Sal	2–10	430	
					21.7–16.8.75	A	2–18	5.3	upper limit
					9–12.1.76	A	2–18	5.3	upper limit
					6.1976	A	2–18	1670	transient
					1975–77	S3	2–11	78	
					1.1977	A	2–18	470	
					6.1978	A	3–6	1060	transient
					6.1978	K	3–9	424	
					14–27.3.79	A	3–6	320–420	transient
19092+076	3U 1912+07 4U 1090+07 A 1908+07	19 09 12 07 37 30	42.09 −0.94	0.1328	1971–73	U	2–6	7.70	−
					1971–73	O7	3–10	87.5	−
					12–25.11.74	A	2–18	22.5	2.5
19094+047	A 1909+04 4U 1908+05 H 1908+050	19 00 24 04 45 00	39.57 −2.32	0.600	1971–73	U	2–6	6.1	
					10–25.11.74	A	2–18	22.5	2.5
					7–13.4.1978	H	2–10	5.5	
19117−108	GX-21	19 11 42 −10 48 50	20.8±0.1 close to 0°		23.6.73	R	4–18	160	transient
19146−589	2A 1914−589 4U 1924−59	19 14 41 −58 57 00	337.92 −26.45	0.2920	1971–73	U	2–6	2.4	−
					1974–76	A	2–18	1.8	−
					10.1977	H	2–10		
19160−793	3U 1849−77 4U 1916−79	19 16 00 −79 18 00	315.07 −28.14	4.5808	1971–73	U	2–6	4.89	−
					1974–76	A	2–18	2.0	upper limit
19161−053	3U 1915−05 A 1916−05 MXB 1916−05 2S 1916−053 4U 1915−05	19 16 08.5 −05 19 51	31.46 −8.48	0.0003	24.9.1970	R	1.2–16	48	
					1971–73	O7	3–10	9	
					1971–73	U	2–6	33	2
					12–25.11.74	A	2.4–19.8	53	1.3
					1975–77	S3	2–11	32	
19185+146	A 1918+14 4U 1918+15	19 18 34 14 37 12	49.34 0.32	0.4000	11.10.1966	R	2–5	300	transient
					1971–73	U	2–6	83	10
					21.7–16.8.76	A	2–18	13.5	6

Table I (continued)

Name KRS	Other names	R.A. decl. (1950)	l b	Area (sq. deg.)	Date	Range keV	Intensity μJy 2–6 keV	Var. comments
1)	(2)	(3)	(4)	(5)	(6)	(7)	(8)	(9)	(10)
19197+436	3U 1921+43	19ʰ19ᵐ46ˢ	75.56	0.0140	1971–73	O7	3–10	7.2	
	2A 1919+438	43°41′36″	13.41		1971–73	U	2–6	7.7	
	4U 1919+44				1974–75	ANS		11.7	
					1974–76	A	2–18	11.7	
					1977–78	H	2–6	4.9	
19202+340	4U 1920+34	19 20 12	66.77	0.1601	1971–73	U	2–6	1.69	–
		34 03 00	9.08						
19262+506	H 1926+503	19 26 12	82.23	0.4000	7.11	H	2–10	1.22	–
		50 19 12	15.22						
19290+079	MX 1929+07	19 29 00	44.60	450	15.1.1975	S3	1–10	1670	burst
		07 54 00	−5.10						
19336+361	4U 1933+36	19 33 36	69.93	0.4271	1971–73	U	2–6	1.54	–
		36 09 00	7.63						
19365+326		19 36 30	67.13	5.6320	7.9.1974	R	0.5–2		–
		32 38 00	5.40						
19386−105	2A 1938−105	19 38 41	29.14	2.240	1974–76	A	2–18	3.2	–
		−10 31 12	−15.77						
19434+364	4U 1943+36	19 43 24	71.12	1.0238	1971–73	U	2–6	3.59	–
		36 24 00	6.03						
19490+440		19 49 00	78.39		23.9.1975	B	20–200		burst
		44 05 00	8.91		4.10.1975				
19540+319	3U 1953+31	19 54 02	68.42	0.0045	1971–73	U	2–6	105	5
	4U 1954+31	31 57 25	1.87		1971–73	O7	3–10	16.7	–
19556−689	3U 1959−69	19 55 36	326.83	0.2947	1971–73	U	2–6	3.51	–
	4U 1955−68	−68 54 00	−31.39		1971–73	O7	3–10	0.5	–
					1974–76	A	2–18	1.7	upper limit
19560+650	1M 1956+65	19 56 00	97.82	2.8000	1971–73	O7	3–10	4.2	–
		65 00 00	18.00						
19564+350	Cyg X-1	19 56 28.8	71.32	optical	1971–73	U	2–6	1960	5
	4U 1956+35	35 03 54.5	3.08	data	1971–73	O7	3–10	430	
					11.1974	ANS	1–28	380	var
					5.1975	ANS	1–28	2500	var
					18.2.1976	A	2–18	950	var
					10.1976	S3	2–10	300	
19570+115	3U 1956+11	19 57 02.9	51.31	0.0001	1971–73	O7	3–10	19.7	–
	A 1956+11	11 34 14	−9.83		12–25.11.74	A	2–18	103	2
	2S 1957+115				1975–77	S3	2–11	35	–
	4U 1956+11				1971–73	U	2–6	29.1	–
19572+405	4U 1957+40	19 57 17	76.10	0.0670	1971–73	U	2–6	4.3	
		40 32 24	5.80		1971–73	O7	3–10	6.2	
					1974	ANS	2–6	11.7	
					1975	ANS	2–6	5.6	
					1975, 1976	O8	2–6	4.3	
					11.1977–5.1978	H	2–6	3	upper limit
20014+626	4U 2001+62	20 01 24	95.88	0.5874	1971–73	U	2–6	2.42	–
		62 36 00	16.36		1974–76	A	2–18	2.0	upper limit
20036+643	4U 2003+64	20 03 36	97.63	0.2158	1971–73	U	2–6	3	–
		64 22 12	16.98						
20091−569	2A 2009−569	20 09 07	340.90	0.0720	1974–76	A	2–18	5.3	–
		−56 57 00	−33.48						
20171+368		20 17 06	75.59		1971–73	U	2–6	9.2	–
		36 51 00	0.85						
20190+395	4U 2019+39	20 19 00	77.50	0.0972	1971–73	U	2–6	5.24	–
		39 30 00	1.75						
20200+405	Cyg X-7	20 20 00	78.50	2.3333	7.9.1974	R	0.5–2		soft, SNR?
		40 30 00	2.20						
20288+428	4U 2028+42	20 28 48	81.28	0.3638	1971–73	U	2–6	5.01	–
		42 49 12	2.18						

Table I (continued)

Name XRS	Other names	R.A. decl. (1950)	l b	Area (sq. deg.)	Date		Range keV	Intensity μJy 2–6 keV	Var. comments
(1)	(2)	(3)	(4)	(5)	(6)	(7)	(8)	(9)	(10)
20305+407	Cyg X-3 4U 2030+40	20ʰ30ᵐ33ˢ 40°47′06″	79°84 0°71	0.000 8	25.4.1965	R	1.5–8	350	–
					11.10.1966	R	2–5	200	–
					25.8.1967	R	2–10	270	–
					3.11.1968	R	2–10	584	–
					5.12.1968	R	1–10	67	–
					24.9.1970	R	1.2–16	200	–
					25.5.1971	R	2–18	150	–
					1971–73	U	2–6	640	2
					1971–73	O7	3–10	130	
					7.9.1974	R	2–6	100	
					3.10.1974	R	2–6	50	
					16–22.11.74	ANS	1.3–7.1	340	
					14–23.5.75	ANS	1.3–7.1	200	
					5.75, 10.76	O8	1.5–6	45	2
					1.1978	S3	1.5–6	100	
20312+317	A 2031+31	20 31 12 31 42 00	72.62 −4.82	0.627 0	7–8.1975	A	2–18	10.3	–
20332+362	A 2033+36	20 33 12 36 15 36	76.53 −2.42	0.160 0	7–8.1975	A	2–18	14.4	–
20380+290	Vul XR-1	20 38 00 29 00 00	71.33 −7.61	7.065 0	25.4.1965	R	4–8	568	–
					7.7.1967	R	0.25		soft
					7.9.1967	R	0.5–13		soft
20407−115	2A 2040−115	20 40 46 −11 33 36	35.21 −29.98	0.398 0	1974–76	A	5.3	4.0	–
20419+754	1M 2041+75	20 41 55 75 25 12	109.36 19.86	1.200 0	1971–73	O7	3–10	2.5	–
20467+319	4U 2046+31	20 46 46 31 54 00	74.81 −7.31	2.464 0	1971–73	U	2–6	0.82	–
20486+443	4U 2048+44	20 48 36 44 22 30	84.71 0.32	0.122 3	1971–73	U	2–6	1.10	–
20497+308		20 49 45 30 53 00	74.20 −8.30		26.6.1970	R	0.5–2		soft, SNR
					30.3.1973	R	0.15–1.8		–
20560+493	3U 2052+47 4U 2056+49	20 56 00 49 19 30	89.30 2.55	0.033 6	1971–73 1971–73	O7 U	3–10 2–6	0.2 5	– –
20582+328	4U 2058+32	20 58 12 32 52 30	77.12 −8.55	4.106 4	1971–73	U	2–6	2.2	–
20586+417	H 2058+417	20 58 36 41 43 48	83.84 −2.79	0.33	30.11.1977	H	2–10	1.6	
21040+315	4U 2104+31	21 04 00 31 30 00	76.89 −10.38	2.604 0	1971–73	U	2–6	1.20	–
21091+385	Cyg X-4	21 09 11 38 33 00	82.90 −6.40	1.000 0	25.4.1965	R	1.5–8	250	transient?
					11.10.1966	R	2–5	80	–
					5.12.1968	R	1–10	30	–
21200+450	Cyg X-6	21 20 00 45 00 00	89.00 −3.40	60.000 0	5.12.1969	R	0.5–1.4		soft
					7.9.1974	R	0.5–2		soft
21203+321	4U 2120+32	21 20 22 32 07 48	79.76 −12.50	2.283 8	1971–73	U	2–6	1.0	–
21275+119	3U 2131+11 2A 2127+120 4U 2129+12 2S 2127+119	21 27 34.2 11 56 51	65.09 −27.19	0.000 1	1971–73 1971–73 1974–76 12.1975	U O7 A S3	2–6 3–10 2–18 2–11	6.41 27 19.4 9.0	– 2 3.6 –
21288+816	1M 2128+81	21 28 48 81 36 00	116.07 21.84	1.100 0	1971–73	O7	3–10	2.1	–
21296+471	4U 2129+47	21 29 34 47 04 29	91.60 −3.01	0.000 2	1971–73 1971–73 9–17.12.77 7.1978	U O7 H	2–6 3–10	33 20.4	2 –
21346+557	4U 0134+55	21 34 36 55 45 00	98.00 2.86	0.197 6	1971–73	U	2–6	3.69	–

Table I (continued)

Name XRS	Other names	R.A. decl. (1950)	l b	Area (sq. deg.)	Date		Range keV	Intensity μJy 2–6 keV	Var. comments
(1)	(2)	(3)	(4)	(5)	(6)	(7)	(8)	(9)	(10)
21370 + 567	Cep X-4	$21^h37^m00^s$	99.°00	0.4220	12.1971	O7	7–26	50	upper limit
		56°47′00″	3.°40		6.1972	O7	7–26	3340	transient
					1971–73	U	2–6	4.8	–
					1971–73	O7	3–10	2.5	
					2.1976	O7	7–26	130	upper limit
21407 + 433		21 40 42.6	90.6	optical	20.6.1971	U	2–6	2	–
		43 21 51	−7.1	data	30.3.1973	R	0.15–0.28		flare
					12.1974	ANS	1–7	8.3	flare
					1975	S3	0.15–0.28		flare
					1974–76	A	2–18	13.6	–
					12.1977	H	2–6	7.7	–
					14.6.1978	H	2–6	9	2
21409 − 602	4U 2126 − 60	21 40 55	333.64	1.3120	1971–73	U	2–6	3.94	–
		− 60 12 00	−44.44		1971–73	O7	3–10	5.0	–
21426 + 380	4U 2142 + 38	21 42 36.91	87.32	optical	1971–73	U	2–6	910	2
	Cyg X-2	38 05 27.9	−11.32	data	1971–73	O7	3–10	520	–
	2A 2142 + 381				1974–76	A	2–18	1350	3.5
					5.7.1975	Sal	2–10	170	–
					1973–77	C	2.5–7.5		
21511 + 309	1M 2151 + 30	21 51 10	74.66	0.3900	1971–73	O7	3–10	3.4	–
		30 56 24	−8.65						
21518 − 316	2A 2151 − 316	21 51 48	15.64	0.2540	1974–76	A	2–18	4.0	–
		− 31 40 12	−51.48						
21541 − 303	H 2154 − 304	21 54 09.6	17.67	2.0400	13.11.1977	H	2–10	4.30	var
		− 30 23 00	−51.85						
21554 − 609	2A 2155 − 609	21 55 24	331.49	0.5150	1974–76	A	2–18	3.7	–
		− 60 57 36	−45.67						
21559 − 304	2155 − 304	21 55 57.9	18.16	?	11–15.11.77	H	3–6	10	3
	H 2155 − 304	− 30 27 54	−52.22		12–15.11.78	H	2–12	5.9	2
22046 + 452		22 04 38	95.13		19–22.12.77	H	0.2–2.8		–
		45 15 06	−8.30						
22049 + 472		22 04 54	96.42		19–22.12.77	H	0.2–2.8		
		47 14 00	−6.78						
22063 + 544	A 2204 + 54	22 06 18	100.70	0.0711	1971–73	U	2–6	4.89	–
	3U 2208 + 54	54 24 00	−1.02		1971–73	O7	3–10	2.5	–
	4U 2206 + 54				7–8.1975	A	2–18	5.0	–
22092 + 261	4U 2209 + 26	22 09 12	83.81	1.0787	1971–73	U	2–6	1.3	–
		26 06 00	−24.13						
22095 − 471	H 2209 − 471	22 09 31.2	349.71	1.39	9.11.1977	U	2–10	1.1	–
		− 47 09 00	−53.21						
22116 + 124		22 11 36	74.04			A	2–18	1.3	–
		12 27 10.8	−34.84						
22139 + 239	4U 2213 + 23	22 13 54	83.22	2.3201	1971–73	U	2–6	0.55	–
		23 54 00	−26.51						
22154 − 086	H 2215 − 086	22 15 28.80	52.95	0.7100	25.11.1977	H	2–10	1.50	–
		− 08 41 12	−49.23						
22204 − 022	2A 2220 − 022	22 20 26	61.84	0.1740	1974–76	A	2–18	4.5	–
		− 02 13 12	−46.47		1977	H	2–10		var?
22248 − 782	4U 2224 − 78	22 24 48	311.59	0.9453	1971–73	U	2–6	2.80	–
		− 78 15 00	−36.64						
22264 + 014	H 2226 + 014	22 26 28.8	67.24	3.1400	2.12.1977	U	2–10	1.1	–
		+ 01 26 24	−45.22						
22330 − 378	H 2233 − 378	22 33 04.8	4.07	1.27	18.11.1977	H	2–10	1.17	–
		− 37 52 12	−59.73						
22373 − 256	2A 2237 − 256	22 37 19	28.85	0.4080	1974–76	A	2–18	2.3	–
	MX 2244 − 24(?)	− 25 37 12	−60.57						
22389 + 607	3U 2233 + 59	22 38 54	107.75	0.2079	1971–73	U	2–6	4.7	–
	4U 2238 + 60	60 43 30	2.03		1971–73	O7	3–10	0.2	–

Table I (continued)

Name XRS	Other names	R.A. decl. (1950)	l b	Area (sq. deg.)	Date		Range keV	Intensity μJy 2–6 keV	Var. comments
(1)	(2)	(3)	(4)	(5)	(6)	(7)	(8)	(9)	(10)
22404+267	4U 2240+26	22^{h}40^{m}24^s 26°42′00″	90°53 −27°69	1.9927	1971–73	U	2–6	3	−
22444−242	MX 2244−24	22 44 24 −24 12 00	32.38 −61.84	0.9030	12.70–6.71 1971–73	U O7	2–6 3–10	5.5 27.0	upper limit
22468+601	A 2246+60	22 46 48 60 06 36	108.32 1.03	1.1760	7–8.1975	A	2–18	5.3	−
22514−178	2A 2251−179	22 51 25.4 −17 50 40	46.21 −61.33	0.0004	1974–76 11–13.8.77	A S3	2–18 2–11	7.7 1.7	8 −
22525+181	4U 2252+18	22 52 34 18 09 00	87.94 −36.42	1.3627	1971–73	U	2–6	0.73	−
22528−035	H 2252−035	22 52 48 −03 31 12	68.61 −53.41	0.3200	6.12.1977	H	2–10	2.36	−
22591+161	4U 2259+16	22 59 07 16 06 00	88.24 −38.99	1.2057	1971–73 1974–76	U A	2–6 2–18	1.30 0.49	− −
22595+085	2A 2259+085 4U 2300+08	22 59 31 08 03 00	82.66 −45.38	0.9200	1974–76 1971–73	A U	2–18 2–6	5.0 3.71	− −
23022−088	2A 2302−088 4U 2305−07	23 02 12 −08 53 24	64.21 −58.73	0.0800	1971–73 1974–76	U A	2–6 2–18	3.3 5.0	− −
23086+597	H 2309+59	23 08 36 59 43 00	134.73 56.49	0.5000	11–20.1.78	H	1–10	4.7	−
23120−214	A 2312−21	23 12 00 −21 24 00	43.02 −67.15	2.0000	1974–76	A	2–18	0.6	−
23153−428	2A 2315−428	23 15 19 −42 48 36	347.81 −65.56	0.0670	1974–76	A	2–18	6.3	5 −
23166+618	4U 2316+61	23 16 36 61 48 00	112.31 1.13	0.2537	1971–73	U	2–6	4	−
23184−272	2A 2318−272 MX 2321−23(?)	23 18 26 −27 12 00	28.32 −69.91	0.5580	1974–76	A	2–18	1.8	−
23210−230	MX 2321−23	23 21 00 −23 00 00	38.00 −70.00	11.6	1971–73 12.70–6.71	O7 U	3–10 2–6	3.0 6.7	upper limit
23212+585	4U 2321+58 1M 2321+58	23 21 13 58 33 29	111.75 −2.12	0.0004	1971–73 1971–73	U O7	2–6 3–10	89.2 49.1	SNR −
23221+166	2A 2322+166 4U 2315+15	23 22 10 16 40 48	94.93 −41.15	0.129	1971–73 1974–76 1975–77	U A O8	2–6 2–18 2–20	3.16 2.3	− · − −
23357+427	4U 2335+42	23 35 44 42 43 30	108.99 −17.88	0.1713	1971–73	U	2–6	1.40	−
23440+086	4U 2344+08	23 44 05 08 39 00	97.15 −50.67	1.7593	1971–73 1974–76	U A	2–6 2–18	4.41 2.7	− upper limit
23448−285	2A 2344−285 4U 2344−27	23 44 50 −28 34 48	24.49 −75.80	0.142	1971–73 1974–76	U A	2–6 2–18	3.09 3.2	− −
23454+273	3U 2346+26 4U 2345+27	23 45 24 27 18 00	106.08 −33.21	0.4430	1971–73 1971–73 1974–76	U O7 A	2–6 3–10 2–18	3.94 0.7 2.8	− − upper limit
23512+067	4U 2351+06	23 51 17 06 46 12	98.67 −53.09	2.2104	1971–73 1974–76 17–22.8.77	U A S3	2–6 2–18 2–10	1.89 2.2 1.36	− upper limit −
23587+210	4U 2358+21	23 58 42 21 04 30	107.62 −40.02	0.2520	1971–73	U	2–6	2.92	burst
23588−640	MX 2348−65	23 58 50 −64 05 00	313.50 −51.60	0.0706	16.11.1975 16.11.1975	S3 A	1.5–10 2–18	330 230	burst burst
B 1	Great region at α ≈ 22 40 − 0 40				11.10.1976	S3	2–11		bursts
B 2	Great region at α ≈ 23 40 − 0 50 δ ≈ −23° to −45°				27.9.1975	A	2–18		burst
B 3	Great region at α ≈ 14 − 15 , δ = −40°				11.10.1975	A	2–18		burst
B 4		02 46 00 23 30 00		120	1973	R	0.2–1		burst

TABLE II

Notes

00000 + 278   Two bursts of duration ~ 2 and ~ 5 hr.
              (Cooke, 1976).

00000 + 726   Forman et al. (1978).

00012 − 310   Possibly it is identical with XRS 00092 − 339.
              (Markert et al., 1978).

00058 + 200   Source possibly confused.
              Optics: Seyfert galaxy Mrk 335.
              (Forman et al., 1978).

00079 + 106   X-rays: α (2–10 keV) $= -0.3 \pm 0.3$; L_x (2–10 keV) $\approx 2 \times 10^{45}$ erg s^{-1}.
              Optics: Quasar IIIZW^2, $V = 14.80$, $A_v = 0.23$, $M_v = -24.1$, $D = 535$ Mpc, $L \sim 10^{45}$ erg s^{-1}.
              Infrared: $\lambda = 1.25\,\mu$m, $S_v = 9.3 \pm 1\,\mu$Jy, $\alpha = -0.4$, $L \sim 2 \times 10^{45}$ erg s^{-1}.
              (Bradt, 1979).

00092 − 339   Optics: Cluster of galaxies SG 000−303 ($L_x = 2.6 \times 10^{45}$ erg s^{-1}), SC 0002−308
              ($L_x = 6.8 \times 10^{44}$ erg s^{-1}), CA 0007−306 ($L_x = 1.5 \times 10^{45}$ erg s^{-1}).
              Radio: Sources PKS 0003 − 302, PKS 0008 − 300.
              (Forman et al., 1978; Lugger, 1978; Melnick and Quintana, 1975; Rowan-Robinson and
              Fabian, 1975).

00108 + 396   Forman et al. (1978).

00123 − 052   Markert et al. (1978).

00153 + 028   Forman et al. (1978).

00218 − 420   Possibly it is identical with the source 00390 + 411.
              (Market et al., 1978).

00224 + 638   SNR Tycho (1572) 3CIO.
              X-rays: L_x (1–10 keV) $= 7 \times 10^{34}$ erg s^{-1}, $N_H \sim 1.5 \times 10^{22}$ cm^{-2}, $n_0 \sim 1.6$ cm^{-2}, $V_r \sim 3000$
              km s^{-1}, $E_0 \sim 2 \times 10^{51}$ erg, $kT = 2.7$ keV.
              Optics: $R = 3.2$ pc, $D = 3$ kpc, age 400 yr.
              Radio: Flux 60 units for 1 GHz, $\alpha = -0.6$, diam. 7.0 pc, distance 3–4 kpc.
              (Culhane, 1974; Gorenstein and Taker, 1975; Markert et al., 1978; Winkler, 1978; Woltjer,
              1972).

00260 − 097   Markert et al. (1978).

00262 + 593   Carpenter et al. (1977); Forman et al. (1978).

00264 − 730   Identified with globular cluster Kron$_3$ (?).
              X-rays: $L_x \approx 10^{37}$ erg s^{-1}.
              Optics: $V = 12.03$, radius of core $\sim 17''$ (~ 5 pc), central density $\rho_0 \sim 10^2\,M_\odot$ pc^{-3}, distance
              ~ 66 kpc, $A_v \approx 0.1$.
              (Forman et al., 1978; Grindlay, 1978).

00268 − 291   Forman et al. (1978).

00288 + 220   Cooke et al. (1978), *Forman et al. (1978)*.

00328 + 242   Markert et al. (1977).

00332 + 588   3C 14.1 (?), (Forman et al., 1978).

00390 + 411   Galaxy M 31.
              X-rays: HEAO-B in 1979 has divided the galaxy into 69 sources, from which 36 sources belong
              to Population I sources, 18 buldge sources, 7 globular clusters and 18 candidates.
              (Cooke et al., 1978; Forman et al., 1978; van Speybroeck, 1979).

00397 − 096   Possibly cluster of galaxies Abell 85 (??).
              X-rays: $L_x = 9.0 \times 10^{44}$ erg s^{-1}.
              Optics: class I, $D_{max} = 411$ mpc, dimension of cluster $0.35^{+0.07}_{-0.04}$ mpc. $L_{opt.} \sim 6 \times 10^{38}$ erg s^{-1}.
              (Cooke et al., 1978; Forman et al., 1978; Murray and Ulmer, 1976; Rowan-Robinson and
              Fabian, 1975; Schwartz, 1978; Schwartz et al., 1979).

00419 + 368   Forman et al. (1978).

00421 + 327   Transient source.
              X-rays: time of burst ~ 2 weeks. There are no pulsations (16 ms–500 s). Spectrum flat in the
              range 1–19 keV. Binary system (?) $P_{orb.} \simeq 11.6$ days, duration of eclipse 2.3 days. $M_{opt.} = 1 M_\odot$,
              $M_x < 0.15\,M_\odot$, $L_x/L_{opt.} = 1800$.

Table II (continued)

	Optics: star 19^m, $\Delta m \simeq 0.4$–0.5, type G0-G8, $B - V = 0.58$ A $(V) \sim 0.7$.
	Radio: flux 20 mJy (610 GHz).
	(Bahcall *et al.*, 1976; Bradt, 1978; Bradt *et al.*, 1978; Markert *et al.*, 1976a; Markert *et al.*, 1977; Murray and Ulmer 1976; Rappaport *et al.*, 1977a; Ricketts and Cooke, 1977).
$00500 + 592$	Markert *et al.* (1977).
$00503 - 727$	Transient source in SMC.
	X-rays: Spectral index $\alpha = -0.99 \pm 0.10$, $L_x(\text{max}) \simeq 6 \times 10^{37}$ erg s^{-1}.
	Optics: $V \sim 15$, $B - V = -0.3$, $U - B = -1.0$, spectrum Be (O9 III-V) distance 63 kpc.
	(Allen, 1978; Bradt *et al.*, 1978; Clark *et al.*, 1978; Li *et al.*, 1977a).
$00520 - 687$	Forman *et al.* (1978).
$00529 - 739$	Transient source in SMC.
	X-rays: Spectral index $\alpha = -0.89 \pm 0.09$, $L_x(\text{max}) = 10^{38}$ erg s^{-1}.
	Optics: $V = 16.0$, $B - V = -0.3$, $U - B = -0.5$, type B1e-B2e, $M_v = -5.0$, $P_{\text{orb.}} \sim 1.5$–4 days (?), total mass 10–25 $M_\odot$.
	(Allen, 1978; Bradt *et al.*, 1978; Clark *et al.*, 1978; Li *et al.*, 1977a; Murdin *et al.*, 1978; Schlosser and van Paradijs, 1979).
$00536 + 604$	Star γ Cas = SAO 011482.
	X-rays: $L_x = 3 \times 10^{33}$ erg s^{-1}, at $D = 300$ pc.
	Optics: $V = 1.6$–3.0, I = 2.2, $K = 2.42$; class BO5e (II-V), $P_{\text{orb.}} = 0.7$ days, $A(V) = 0.16$, $M_v = -4.5$.
	Radio: Flux <10 mJy (2695 GHz).
	(Bradt *et al.*, 1977a; Bradt *et al.*, 1978; Cowley *et al.*, 1976b; Ferrari-Toniolo *et al.*, 1978; Jernigan, 1976a; Markert *et al.*, 1977).
$00549 - 015$	Cluster of galaxies Abell 119. Distance 312 Mpc, $L_x \sim 5 \times 10^{44}$ erg s^{-1}.
	(Cooke *et al.*, 1978; McHardy, 1978; Schwartz, 1978).
$00554 - 796$	Markert *et al.* (1977).
$00577 - 239$	Markert *et al.* (1977).
$01020 - 242$	Cluster of galaxies Abell 140, 141 (?).
	(Cooke *et al.*, 1978; McHardy, 1978).
$01039 - 218$	Cluster of galaxies Abell 133 (?). Distance 449 Mpc, $L_x = 10^{45}$ erg s^{-1}.
	(Cooke *et al.*, 1978; Forman *et al.*, 1978; McHardy 1978; Schwartz, 1978).
$01114 - 149$	Cluster of galaxies Abell 151 (?).
	(Marshall *et al.*, 1979).
$01147 + 650$	Optics: Star LSI$+65°010$, $\alpha = 1^h14^m41^s8$, $\delta = 65°01'32''$, $V \approx$ II, type B0.5 IIIe, $B - V = 1.2$, $U - B = 0.1$, $A_v = 5.0$, $L_x/L_{\text{opt.}} = 1.5 \times 10^{-4}$.
	(Bidelman and Sanduleak, 1978; Bradt *et al.*, 1978; Dower and Kelley, 1977; Margon and Bradt, 1977).
$01152 + 634$	Transient source.
	X-rays: Spectrum $kT > 15$ keV, $L_{\text{max}} \sim 3 \times 10^{37}$ erg s^{-1}. $P_{\text{puls.}} = 3.614\,573\,7$ s, $P_{\text{orb.}} = 24.309$ days, $\dot{P}_{\text{puls.}}/P_{\text{puls.}} = (-3.2 \pm 0.8)\ 10^{-5}$ yr^{-1}. The absorption line on energy 20.1 ± 0.5 keV, $\tau \sim 14$–30 days, $f(M) = 5M_\odot$, $a_x \sin i = 4.2 \times 10^{12}$ cm, $K_x = 133.65$ km s^{-1}, $e = 0.3402$.
	Optics: Star $\alpha = 1^h15^m13^s8$, $\delta = 63°28'38''$, $V = 15.64$, $B - V = 1.44$; $U - B = 0.31$; $A_v \geqslant 5$, type Be, $M_v \geqslant -4$, distance 5–7 kpc, $M_{\text{opt.}} > 5M_\odot$.
	(Bradt *et al.*, 1978; Cominsky *et al.*, 1978 a, b; Griffiths *et al.*, 1978; Jones *et al.*, 1978; Johnson *et al.*, 1978; Rappaport *et al.*, 1978, 1978a).
$01157 - 737$	Source in SMC.
	X-rays: binary system $P_{\text{orb.}} = 3.892\,29$ days, pulsar $P_{\text{puls.}} = 0.714\,733\,7$ s, $\dot{P}/P = (-5.01 \pm 0.09)\ 10^{-4}$ yr^{-1}, $e \approx 0.0003$, $K_x = 300.5$ km s^{-1}, $N_H = 3.4 \times 10^{20}$ cm^{-2}, $M_x = 1.1$–$4M_\odot$, $f(M) = 10.8M_\odot$, $a_x \sin i = 1.6 \times 10^{12}$ cm, $L_x \sim (5$–$13) \times 10^{38}$ erg s^{-1}.
	Optics: Binary system Sc 160, $V = 13.2$, $\Delta V = 0.09$, spectrum BO1, $M_v = -6.0$, $K_{\text{opt.}} = 40$ km s^{-1}, $M_{\text{opt.}} = 16.2 \pm 1.4M_\odot$.
	(Avni and Bahcall, 1975; Bradt *et al.*, 1978; Bunner and Sanders, 1979; Clark *et al.*, 1977; Cooke *et al.*, 1978; Forman *et al.*, 1978; Gursky and Schreier, 1975; Henry and Schreier, 1977; Hutchings, 1978; Jentis *et al.*, 1977; Lucke *et al.*, 1975; Schreier, 1977).

Table II (continued)

01206−591	Seyfert galaxy F-9 (?).
	X-rays: $L_x \simeq 2 \times 10^{44}$ erg s^{-1}.
	Optics: $M = 10.63 \pm 0.23$, $K = 10.37 \pm 0.12$, $L = 9.47 \pm 0.20$, $m \sim 13$, $V = 13.23$, $A_v = 0.21$,
	$M_v = -24.2$, distance 275 Mpc, $L_{\text{opt.}} \sim 1 \times 10^{45}$ erg s^{-1} (0.3–0.7 μm). Infrared: $\lambda = 2.2\,\mu$m,
	$S_v = 46 \pm 5$ mJy, $\alpha = -1.0$, $L \sim 2 \times 10^{45}$ erg s^{-1}.
	(Bradt, 1979; Cooke *et al.*, 1978; Danks *et al.*, 1979; Forman *et al.*, 1978; Ricker, 1977).
01209−353	X-rays: Luminosity (3–6 keV) $\times 10^{43}$ erg s^{-1}.
	Optics: Emission galaxy NGC 526a, $Z = 0.018$.
	(Cooke *et al.*, 1978; Forman *et al.*, 1978; Griffiths *et al.*, 1979).
01224+338	Cooke *et al.* (1978).
01232+075	Marshall *et al.* (1979).
01296−099	Forman *et al.* (1978).
01322+007	Transient source.
	(Barnden and Francey, 1969; Kaluzienski, 1977).
01345−115	Forman (1978), Ricketts (1978).
01364−182	Flare star UV Cet.
	X-rays: duration of burst ~ 3 min (8.1.75), maximum 7.3×10^{-11} erg cm^{-2} s^{-1} (0.2–0.28 keV).
	Energy of burst 3×10^{30} ergs.
	Optics: Flare synchronous with X-rays, $\Delta m \geqslant 6$, in the maximum luminosity (in colour U)
	$>2 \times 10^{30}$ erg s^{-1}. Star is of dM 5.5e type, $V = 6$.
	(Heise *et al.*, 1975; Tsikoudi and Hudson, 1975).
01385+480	Forman (1978).
01427+612	Optical candidate star $V \sim 20$.
	(Bradt *et al.*, 1978; Forman *et al.*, 1978; Markert *et al.*, 1977).
01430−330	Source existance is doubtful. Identification with cluster SC 0141−340 ($\alpha = 01^{\text{h}}14^{\text{m}}18^{\text{s}}$,
	$\delta = -34°05'$) is possible. Then luminosity $L_x = 1.2 \times 10^{45}$ erg s^{-1}.
	Radio: source PKS 0141−344.
	Lugger (1978).
01486+360	Cluster of galaxies Abell 262 (?). Galaxy Markaryan 2 (?).
	X-rays: $kT = 2.6 \pm 0.4$ keV.
	(Cooke *et al.*, 1978; Forman *et al.*, 1978; Henry and Tucker, 1979; Ricketts, 1978).
02065−019	Galaxy MKN 590 (?).
	(Marshall *et al.*, 1979).
02140+045	Transient source.
	X-rays: Bremsstrahlung spectrum, $T = 8.3 \times 10^{7}$ K.
	(Shulka and Wilson, 1970).
02140+623	Supernova remnant HB 3.
	X-rays: flux 6.4×10^{-11} erg cm^{-2} s^{-1} (0.6–2.2 keV).
	Radio: diam 80 pc, flux 36 mJy (1-GHz), spectral index -0.7, distance 1.7 kpc.
	(Galas *et al.*, 1978).
02202+184	NGC 0918 (?).
02238+312	Forman *et al.* (1978).
02272+437	Possibly, it is XRS 02530+417 (?).
	(Markert *et al.*, 1977). (Marshall *et al.*, 1979).
02285−130	Cluster of galaxies Abell 358 (?), distance 545 Mpc, luminosity $L_x = 1.5 \times 10^{45}$ erg s^{-1}.
	(Forman *et al.*, 1978; Ricketts, 1978; Schwartz *et al.*, 1978).
02352−526	Cooke *et al.* (1978).
02410+622	Quasar (?).
	X-rays: flux (300–1200 keV) $\approx 2.3 \times 10^{-5}$ phot. cm^{-2} s^{-1} (Ariel −5, 7.75). In 2–10 keV
	$L_x \sim 3 \times 10^{44}$ erg s^{-1}, spectral index $\alpha = -0.8 \pm 0.1$, source associates with gamma-source
	CG 176–7.
	Optics: quasar or seyfert galaxy, $V \sim 15.7$–16.4, $A_v = 4.6 \pm 1.7$, $M_v = -25.8 \pm 1.8$, distance
	~ 250 Mpc, luminosity $\sim 4 \times 10^{45}$ erg s^{-1}.
	Radio: flux 290 mJy on 1.5 GHz, luminosity (1–90 GHz) $\sim 2.5 \times 10^{42}$ erg s^{-1}, $\alpha = 2^{\text{h}}41^{\text{m}}00^{\text{s}}68$,
	$\delta = 62°15'27''5$.

Table II (continued)

	(Apparao *et al.*, 1978a; Apparao *et al.*, 1978; Bradt, 1979; Coe *et al.*, 1976c; Forman *et al.*, 1978; Julien and Helmken, 1978; Maraschi *et al.*, 1977).
02484−853	Forman *et al.* (1978).
02522+060	Cluster of galaxies Abell 400 (?).

02484−853 Forman *et al.* (1978).

02522+060 Cluster of galaxies Abell 400 (?).
X-ray luminosity (2–10 keV) 5×10^{43} erg s^{-1}.
(Cooke *et al.*, 1978; McHardy, 1978).

02528+440 Marshall *et al.* (1979).

02530+417 Possibly, it is X-ray source 02272+437. Cluster of galaxies Abell 347/396 (?).
X-rays: $kT = 4.1$ (+0.8–0.7) keV, $N_{\rm H} < 1.3 \times 10^{21}$ cm^{-2}.
(Cooke *et al.*, 1978; Markert, 1976a; Markert *et al.*, 1977; Mushotzky *et al.*, 1978; Murray and Ulmer, 1976).

02538+193 Seyfert galaxy Markaryan 372.
Marshall *et al.* (1979).

02554+132 Cluster of galaxies Abell 401 (?).
X-rays: $kT = 7.6 \pm 0.6$ keV, $N_{\rm H} = (1\text{–}24)\ 10^{21}$ cm^{-2}, luminosity (1–10 keV) $= 1.9 \times 10^{45}$ erg s^{-1}. Extended source, dimension $25\overset{'}{.}5 \pm 4\overset{'}{.}4$.
Optics: distance ~ 530 Mpc.
(Cooke *et al.*, 1978; Forman *et al.*, 1978; Henry and Tucker, 1979; Markert *et al.*, 1978; Mushotzky *et al.*, 1978).

02586+607 Markert *et al.* (1977).

03029−223 Forman *et al.* (1978).

03049+407 Eclipsing star β Per (Algol).
X-rays: in the range 0.15–2.7 keV intensity 1.2×10^{-10} erg cm^{-2} s^{-1}, $T \simeq 3 \times 10^{7}$ K, luminosity (2–6 keV) 2×10^{31} erg s^{-1}.
Optics: $V = 2.12 - 2.2$, $\Delta V = 1.3$, distance 30 pc, $P_{\rm orb.} = 2.867$ days. System is triple; spectra B8 V; K0 IV, A5 IV, $B - V = -0.1$, $M_v = -0.3$.
Radio: variable nonthermal source, dimension 0.02a, e., flares up to 1 Jy, on 8 GHz.
(Bradt *et al.*, 1978; Brini *et al.*, 1976; Epstein, 1975; Riegler *et al.*, 1978; Schnopper *et al.*, 1976).

03059+530 Markert *et al.* (1978), Forman *et al.* (1978).

03064+477 LX Per, system of RS CVn type.
X-rays: $L_x = (2.2 \pm 0.2)\ 10^{31}$ erg s^{-1} (0.2–2.8 keV) if the distance 145 pc.
Optics: G4, $m = 8.4$–9.4, $P = 8.038044$ days, $M - m = 7$, distance 145 pc.
(Waltjer *et al.*, 1978).

03102+465 Forman *et al.* (1978).
At distance 25 pc $L_x \simeq 10^{33}$ erg s^{-1} (20–40 keV).
Optics: in the box cluster of galaxies Per A, planetary nebulae IC 351 and 2C 2003, close binary β Per are located.
Radio: in the box remnant CTB 13, some radioglalaxies are located.
(Fuligni *et al.*, 1976).

03110+420 X-rays: 15 bursts, spectrum is harder than Crab.
Duration: $2 < t < 128$ s.
(Belian *et al.*, 1976).

03116−227 X-rays: flux in the range of 0.15–0.28 keV (SAS-3) 0.91 μJy, $T \sim 10^{7}$ K. Period 91.0 ± 0.1 min.
Optics: star of type AM Her $\alpha = 03^{\rm h}12^{\rm m}00\overset{s}{.}0$, $\delta = -22°46'48''$, $P_{\rm orb.} = 81.04 \pm 0.01$ min, $V = 13.8$–15.0, $B - V = 0.9$, $K = 355$ km s^{-1}, strong emission Hα and He II, optical bursts, polarization.
(Charles *et al.*, 1978; Cooke *et al.*, 1978; Griffiths *et al.*, 1978; Hearn, 1978; Hiltner, 1978; Ward *et al.*, 1979; Williams *et al.*, 1979).

03118+530 Markert *et al.* (1978), Forman *et al.* (1978).

03129+345 Star HD 20210.
X-rays: flux (0.16–0.284 keV) 5.5×10^{-11} erg cm^{-2} s^{-1}.
Optics: star of A 2m type, $V = 6.42$, $E (B - V) = 0$, $M_v = 2.6$, distance 53 kpc, $P_{\rm orb.} = 5.543491$ days.
(Den Boggende *et al.*, 1978).

Table II (continued)

03140+380	X-rays: flux in the range 0.2–1 keV 1.57 × 10^{-2} μJy (flare), at distance 13.7 pc luminosity (0.2–1 keV) 3.4 × 10^{31} erg s^{-1}.
	Optics: possibly, star with Hα emission, type dM 1.5e distance 13.7 pc.
	(Lategan, 1978).
01362−443	Cluster of galaxies PKS 0316−44 (?).
	Cluster dimensions 1.1 Mpc.
	(Cooke *et al.*, 1978; McHardy, 1978; Schwartz *et al.*, 1979).
03165+413	Cluster of galaxies Per A, center in NGC 1275.
	X-rays: extended source, $kT \sim$ 6–8 keV. $N_H = (9.2 \pm 0.7)$ 10^{21} cm^{-2}, diam $\sim$ 750 kpc, $L_x \sim$ 5 × 10^{44} erg s^{-1}. Point source, $kT <$ 2 keV, $L_x \sim 10^{44}$ erg s^{-1}.
	(Cooke *et al.*, 1978; Forman *et al.*, 1978; Fritz *et al.*, 1971; Helmken *et al.*, 1978; Henry and Tucker, 1979; Markert *et al.*, 1978; McHardy, 1978; Mushotzky *et al.*, 1978; Schnopper and Delvaille, 1977b; Ulmer and Jernigan, 1978).
03210−450	Cluster of galaxies 3C 0316−458 (?).
	Radio source PKS 0317−456.
	X-rays: luminosity $\sim 10^{45}$ erg s^{-1}.
	(Cooke *et al.*, 1978; Forman *et al.*, 1978; Lugger, 1978).
03210+236	Soft X-ray flare, possibly, connected with flare of star of dM Ge type. Coordinates are given according to optical object. Distance 17.5 pc, intensity in the range 0.2–1 keV is 77.3 μJy (maximum of flare).
	(Lategan, 1978).
03226+595	Forman *et al.* (1978).
03228+285	Star UX Ari, type RS CVn.
	X-rays: $T \sim 10^7$ K, emission Fe XVII on 0.9 keV. Luminosity in the range 0.2–0.8 keV is $L_x =$ 2.1 × 10^{31} erg s^{-1}.
	Optics: $V_{max} =$ 6.49, type G5 V + K0 IV, distance 50 pc, $P_{orb.} =$ 6.437 91 days.
	(Walter *et al.*, 1978a, b).
03250+440	X-rays: duration of flare is 12 min (20–40 keV), $kT =$ 3.5 keV, at distance 73 Mpc $L_x \simeq$ 8 × 10^{48} erg s^{-1}.
03277+432	Optics: Nova GK Per (1901 yr) $\alpha =$ 3^{h}27^{m}45^{s}6, $\delta =$ 43°44′24″, 12.9 $\geqslant M_v \geqslant$ 12.3, max. $\sim$ 1^m. K2 IV-V star, $M_v \sim$ 5.4. Distance $\sim$ 460 pc, $A_v \approx$ 0.2.
	(King *et al.*, 1979; Ricketts *et al.*, 1978).
03330+317	Quasar NRAO 140 (?).
	X-rays: if identification is correct, then luminosity 4 × 10^{47} erg s^{-1}.
	Radio: variable source, flares.
	(Marshall *et al.*, 1979).
03318−363	X-rays: emission galaxy NGC 1365 (?), $M_{pg} =$ 11.2, luminosity (2–10 keV) 1.7 × 10^{42} erg s^{-1}.
	(Griffiths *et al.*, 1979; Ward *et al.*, 1978).
03342−302	Forman *et al.* (1978).
03342+002	Possibly, it is XRS 03362+010 (?).
	X-rays: intensity (0.2–2.8 keV) 0.04 cts cm^{-2} s^{-1}, luminosity (0.2–2.8 keV) (1–3) × 10^{31} erg s^{-1}. Duration of bursts 3 h, synchronous with radioflare.
	Optics: star V 711 Tau (?), $V_{max} =$ 5.9, $P_{orb.} =$ 2.838 days, spectrum G5V + K0V, distance 33 pc, activity Hα, Lα.
	(Walter *et al.*, 1978a, b; 1978; White *et al.*, 1978).
03353+096	Cooke *et al.* (1978), Forman *et al.* (1978).
03362+010	Transient source.
	(Forman *et al.*, 1978).
03385+500	Villa *et al.* (1976).
03432−536	Identification with clusters of galaxies SG 0320−515, CA 0325−539, CA 0329−527, CA 0340 −538 is possible.
	X-ray luminosity (2–15) × 10^{44} erg s^{-1}.
	(Lugger, 1978; Markert *et al.*, 1978; McHardy, 1978; Melnick and Quintana, 1975).
03492−139	Quasar PKS 0349−1404 (??).
	(Cooke *et al.*, 1978).

Table II (continued)

03522+308   X Per

X-rays: pulsar $P = 13.9332 \pm 0.0002$ min. Period $P_{orb.} = 580 \pm 30$ days is also found. $\dot{P}/P = -2.72 \times 10^{-3}$ min yr^{-1}. Spectrum $kT = 10.4 \pm 0.2$ keV (Ariel). Luminosity in the range 2–6 keV is 5×10^{33} erg s^{-1}. $N_H = 2.4 \times 10^{21}$ cm^{-2}.

Optics: star X Per, $V = 6.0$–6.7, type 09.5 (III-V)e, $M_v = -3.6$, $A_v = 1.8$. According to radial velocities data and photometry $P = 580$ days. The line He II 4686 also gives $P = 13.924$ min. (Becker *et al.*, 1979; Bradt, 1978; Bradt *et al.*, 1978; Campisi *et al.*, 1976; Cooke *et al.*, 1978; Dorren *et al.*, 1978; Ferrari-Toniolo *et al.*, 1978; Forman *et al.*, 1978; Frontera *et al.*, 1979; Gottlieb, 1975; Guinan *et al.*, 1977, 1978; Hutchings, 1977; Jones, 1976; Markert *et al.*, 1978; Murdin *et al.*, 1976; Penston *et al.*, 1976; Persi *et al.*, 1977; Robinson and Africano, 1975; White and Mason, 1977).

03531−400   X-rays: duration of each burst ∼5 hr.

Optics: star in cluster NGC 5367 (?), strong emission Hα. No other stellar lines. (Cooke, 1976; Cooke *et al.*, 1978; Soderblom, 1976).

03576−743   Forman *et al.* (1978).

04040+476   Forman *et al.* (1978).

04064−308   Forman *et al.* (1978).

04074+379   Forman *et al.* (1978); Marshall *et al.* (1979).

04106+103   Cluster of galaxies Abell 478 (?).

X-rays: $kT \approx 10$ keV, luminosity ∼2×10^{45} erg s^{-1} (?). (Cooke *et al.*, 1978; Forman *et al.*, 1978; Henry and Tucker, 1979; Markert *et al.*, 1978; McHardy, 1978; Rowan-Robinson, 1975; Schnopper and Delvaille, 1977b; Schwartz *et al.*, 1979).

04150−120   X-rays: intensity (0.18–0.28 keV), 7×10^{-10} erg cm^{-2} s^{-1}, $kT = (0.2$–0.3) keV. This source may be SNR (?). Luminosity (0.18–0.28 keV) (2–15) × 10^{34}erg s^{-1}, distance ∼100–300 pc. diam 17–67 pc, age (1–20) 10^5 yr., $N_H \simeq 6 \times 10^{20}$ cm^{-2}. (Naranan *et al.*, 1976).

04150+379   N-galaxy 3C III (?). Possibly, it is XRS 04074+379 (?).

X-rays: luminosity ∼5.6×10^{44} erg s^{-1} in the range 2–10 keV.

Optics: $V = 18$, $Z = 0.0485$. (Marshall *et al.*, 1978, 1979).

04215+347   Forman *et al.* (1978), Markert *et al.* (1978).

04234−531   Cluster of galaxies SC 0417−558 (?).

X-ray luminosity 2×10^{42} erg s^{-1}. (Cooke *et al.*, 1978; Forman *et al.*, 1978; Lugger, 1978; Markert *et al.*, 1978).

04276−077   Cluster of galaxies Abell 494.

X-rays: luminosity (2–6 keV) ∼ 5×10^{45} erg s^{-1}. (Forman *et al.*, 1978; Ricketts *et al.*, 1978; Schwartz, 1978).

04293−310   Forman *et al.* (1978).

04305+052   Seyfert galaxy 3C 120.

X-rays: luminosity (2–10 keV) (2–4) × 10^{44} erg s^{-1}. (Cooke *et al.*, 1978; Forman *et al.*, 1978; Marshall *et al.*, 1978; Murray and Ulmer, 1976; Rowan-Robinson and Fabian, 1975; Schnopper *et al.*, 1977a; Schnopper and Delvaille, 1977).

04309−615   Cluster of galaxies SG 0427−629 or SC 0430−616. (Lumonosity in the range 2–6 keV is 5×10^{44} erg s^{-1}.) (Cooke *et al.*, 1978; Forman *et al.*, 1978; Lugger, 1978; Markert *et al.*, 1978; McHardy *et al.*, 1978).

04312−136   Cluster of galaxies Abell 496 (?).

X-rays: luminosity (2–6 keV) ∼ 2×10^{44} erg s^{-1}. (Cooke *et al.*, 1978; McHardy, 1978; Schwartz, 1978; Schwartz *et al.*, 1979).

04360+120   Duration of X-ray burst ∼11 s. Burst observed by Uhuru, IMP-6, OGO-5. (Pizzichini *et al.*, 1975).

04400+069   Markert *et al.* (1978).

04432−095   Cooke *et al.* (1978), Forman *et al.* (1978), Markert *et al.* (1978), Ricketts, (1978).

Table II (continued)

04461 + 449	Cluster of galaxies 3C 129.1 (?).
	X-rays: $kT = 5.4\,(+1.0 - 0.7)$ keV, $N_{\mathrm{H}} = 6.1 \times 10^{21}$ cm^{-2}.
	(Cooke *et al.*, 1978; Forman *et al.*, 1978; Markert *et al.*, 1978; Mushotzky *et al.*, 1978; Schwartz
	et al., 1978).
04476 − 037	Marshall *et al.* (1979).
04490 − 550	X-rays: duration of burst ~ 200 min.
	Optics: M-dwarf (?).
	(Bradt *et al.*, 1978).
04495 + 668	Markert *et al.* (1978).
04527 − 742	Marshall *et al.* (1979).
04566 − 449	Possibly, it is source XRS 05124 − 400 (?).
	(Cooke *et al.*, 1978; Forman *et al.*, 1978; Giacconi *et al.*, 1974).
04574 − 357	Group of galaxies (?).
	Forman *et al.* (1978).
05000 − 554	Duration of burst from 1 min up to 2 days.
	Crab-like spectrum. Ea $\leqslant 3$ keV.
	(Kaluzienski *et al.*, 1978).
05032 + 357	Villa *et al.* (1976).
05048 − 843	Forman *et al.* (1978).
05050 − 213	Cluster of galaxies Abell 514 (?).
	X-rays: luminosity in the range 2–6 keV is 6.9×10^{44} erg s^{-1} at distance 370 Mpc.
	(Forman *et al.*, 1978).
05065 − 034	Forman *et al.* (1978).
05096 + 019	Forman *et al.* (1978).
05106 − 446	Markert *et al.* (1978).
05124 − 400	Globular cluster NGC 1851.
	X-rays: rise time of burst ~ 1 s, decay time ~ 10 s, energy of burst $\sim 10^{-7}$ erg cm^{-2}. Interval
	between bursts 0.25 days. Spectrum of stable source $kT \simeq 7.5$ keV.
	Optics: $V = 7.13$, spectrum F7, mass of cluster $1.7 \times 10^5\ M_{\odot}$, mass of the core $2.3 \times 10^4\ M_{\odot}$,
	$V_{\mathrm{esc}} > 30$ km s^{-1}. Distance 9.3 kpc.
	(Bahcall *et al.*, 1977; Clark *et al.*, 1975; Clark and Li, 1977a; Forman and Jones, 1976; Markert
	et al., 1976a, 1978; Pye and Cooke, 1976; Ulmer *et al.*, 1976).
05130 + 459	α Aur (Capella).
	X-rays: luminosity in the range 0.2–2.8 keV is 4×10^{30} erg s^{-1}, spectrum $kT \approx 0.3$–1 keV.
	Optics: spectral binary star, $V = 0.08$, $P_{\mathrm{orb.}} = 104.203$ days, spectra F8–G0 III and G5 III,
	masses 2.9 and 3.0 $M_{\odot}$, $a = 1$ a.e., distance 14 pc.
	(Catura *et al.*, 1975; Walter *et al.*, 1978).
05150 + 384	Forman *et al.* (1978).
05174 − 456	N-galaxy Pic A.
	X-rays: luminosity in the range 2–10 keV is 8×10^{43} erg s^{-1}.
	(Marshall *et al.*, 1978, 1979).
05175 + 175	Peculiar galaxy 3C 138 (?).
	(Forman *et al.*, 1978; Marshall *et al.*, 1979).
05185 − 262	Forman *et al.* (1978).
05198 + 065	Cluster of galaxies Abell 539 (?).
	(Forman *et al.*, 1978).
05212 − 365	Object of BL Lac type PKS 0521 − 365 (?).
	Optics: $B \simeq 16.0$, variability ~ 3.7.
	Radio: flux 20.9 Jy on frequency 960 GHz.
	(Schwartz *et al.*, 1978).
05213 − 719	Source in LMC.
	X-rays: luminosity in the range 3–10 keV is $(1$–$3) \times 10^{38}$ erg s^{-1}, $kT = 9.4 \pm 1.9$ keV,
	$N_{\mathrm{H}} \simeq 8.1 \times 10^{21}$ cm^{-2}.
	Optics: star $B = 17$, $M_v \sim -2$, $\Delta B \sim 0.5$, emission Hα, He II 4686; $L_x/L_{\mathrm{opt.}} \sim 500$.

Table II (continued)

(Cooke *et al.*, 1978; Forman *et al.*, 1978; Markert and Clark, 1975a; Markert *et al.*, 1978; Schnopper and Delvaille, 1977b).

05263−683   Source in LMC.
Spectrum $kT = 1.9 \pm 0.6$ keV, $N_H \sim 3 \times 10^{21}$ cm^{-2}.
(Markert and Clark, 1975a; Markert *et al.*, 1976a).

05263−661   Hard X-ray burst, peak flux 2×10^{-3} erg cm^{-2} s^{-1}. Oscillation period 8.0 ± 0.05 s, average decay time constant 50 s. Three weak events also observed. Total output 4×10^{44} erg, if source in LMC.
(Barat *et al.*, 1979; Cline, 1979; Mazets *et al.*, 1979).

05265−350   Existance of source is doubtful.
Identification with cluster of galaxies SG 0526−357 (?).
X-rays: luminosity (2–6 keV) is 10^{45} erg s^{-1}.
(Lugger, 1978).

05276−328   X-rays: intensity in the range 0.1–3 keV is 1.1×10^{-11} erg cm^{-2} s^{-1}, luminosity (>0.5 keV) 3.4×10^{31} erg s^{-1}, power spectrum, index ~ -1.9, $N_H \simeq 6 \times 10^{22}$ cm^{-2}.
Optics: star $V \sim 14$, emission lines He I, He II, etc.
Distance ~ 100 pc, $L_{\text{opt.}} \approx 10^{32}$ erg s^{-1}.
(Cooke *et al.*, 1978; Charls *et al.*, 1978, 1979; Schwartz *et al.*, 1978, 1979; Thomas *et al.*, 1978).

05303−370   Cluster of galaxies 0527−368 (?).
X-rays: luminosity 4.8×10^{45} erg s^{-1}.
(Cooke *et al.*, 1978; Giacconi *et al.*, 1974; Murray and Ulmer, 1976).

05315+219   Crab nebula = SNR 1054 yr.
X-rays: luminosity $\sim 9 \times 10^{37}$ erg s^{-1}, spectral power index $\alpha = -1.08$, polarization (2.6–5.2 keV) 15–19%. Pulsar $P = 0.033$ s.
Optics: $V_r \sim 1400$ km s^{-1}, $Eo \sim 2 \times 10^{49}$ erg, distance 2 kpc, $A_v \sim 1,5$, mass of envelope $0.1 M_\odot$, pulsar $P = 0.033$ s.
Radio: flux 1000 J on 1 GHz, power spectrum, index, ~ -0.25. Pulsar $P = 0.0331$ s, intensity = some percents of nebulae intensity. Possibly, PSR 0531+21 is connected with gamma source CG 185–5.
(Forman *et al.*, 1978; Reina, 1974; Weisskopf *et al.*, 1976; Woltjer, 1972).

05328−056   X-rays: spectrum $kT = 2.7 \pm 0.8$ keV, $N_H = 2 \times 10^{21}$ cm^{-2}, luminosity (1–8 keV) 4×10^{33} erg s^{-1}.
Optics: $\theta\ ^2$Ori spectral binary, $P_{\text{orb.}} = 20.967\,22$ days, $F(M) = 1.53$, measured magnetic field 1700 G.
(Cooke *et al.*, 1978; Forman *et al.*, 1978; Murray and Ulmer, 1976).

05328−663   Source in LMC.
X-rays: high state of short duration with soft spectrum and bursts (~ 20 s), low state of long duration with hard spectrum. Binary system, $P_{\text{orb.}} = 1.408\,29$ days. X-ray luminosity (2–6 keV) $\sim 2 \times 10^{38}$ erg s^{-1}, mass of X-ray component 2–4 $M_\odot$.
Optics: star $V = 14.0$, $\Delta V = 0.15$, spectrum O7, $k_{\text{opt.}} \simeq 500$ km s^{-1}, He II 4686 emission, mass $\sim 25 M_\odot$, $M_v = -5$, luminosity $\sim 6 \times 10^{38}$ erg s^{-1}.
(Blanco, 1977; Markert and Clark, 1975a; Persch, 1976; Schnopper and Delvaille, 1977b).

05347−581   Marshall *et al.* (1979).

05357−668   X-rays: $kT = 6.5$ keV, $N_H < 2 \times 10^{22}$ cm^{-2}, if X-ray source in LMC, then luminosity (2–10 keV) 8.5×10^{38} erg s^{-1}.
Optics: star $V = 12.8$, $B - V = -0.13$, $U - B = -0.88$, $E_{B-V} = 0.22$
(Johnston *et al.*, 1979; White and Carpenter, 1978).

05357+262   Transient source.
X-rays: flares with the characteristic time ~ 1 mounth and interval ~ 0.5 yr. Spectrum $kT \geqslant 15$ keV, $Ea \leqslant 2.6$ keV.
Pulsar $P = 103.8274 \pm 0.0004$ s, modulation 25%. Luminosity in maximum (2–6 keV) $\sim 2 \times 10^{37}$ erg s^{-1}.
Optics: HDE 245770 star, Bpe spectrum, $V = 9$, $M_v = -6$, $A_v = 2.6$. Strong Hα emission, variable. Distance 1.8–2.5 kpc. Orbital period >49 days.

Table II (continued)

<table>
<tr><td></td><td>(Baratta et al., 1978; Bartolini et al., 1978; Carpenter et al., 1978; Ciatti et al., 1977; Clark and
Chartres, 1978; Cominsky et al., 1978; Forman et al., 1978; Giangrande et al., 1978; Jones, 1976;
Jones et al., 1976b; Joss, 1975; Kaluzienski et al., 1975a; Kaluzienski and Holt, 1978; Markert et
al., 1977; Persi et al., 1979; Rappaport et al., 1976a; Ricker et al., 1975; Ricker and Primini, 1977;
Ricketts et al., 1975a; Rossiger, 1976, 1978; Schmidt and Romanishin, 1975; Soderblom, 1976;
Stier and Miller, 1976; Wade and Oke, 1977; Willmore, 1977).</td></tr>
</table>

05370−441   BL Lac type object PKS 0537−441 (?).
Optics: $M_B < 18.0$, opt. variability 150.
Radio: PKS 0537−441.
(Schwartz *et al.*, 1978).

05382−661   Bar in LMC.
(Rappaport *et al.*, 1975).

05389−641   Source in LMC.
X-rays: strong variability, characteristic time days and hours. Spectrum $kT = 20 \pm 19$ keV,
$N_H \sim 3.5 \times 10^{21}$ cm^{-2}. Luminosity (2–6 keV) 7×10^{37} erg s^{-1}.
Optics: star $V = 16.9$, $B − V = −0.06$, $U − B = −0.66$, spectrum OB III-IV, weak emission of
P Cyg type. Distance 55 kpc.
(Cooke *et al.*, 1978; Forman *et al.*, 1978; Johnston *et al.*, 1978, 1979a; Markert and Clark, 1975a;
Markert *et al.*, 1978).

05390−669   Possibly, it is X-ray source LMC X-4 (?).
X-rays: rise time 1–2 hr, decay time 8 h. Period 80 s is possible.
(White *et al.*, 1977).

05401−697   Source in LMC (SNR?).
X-rays: extended source, dimension 1.3 ± 0.3 (10 pc). Spectrum $kT = 2.7 \pm 0.2$ keV,
$N_H = 1.1 \times 10^{21}$ cm^{-2}, luminosity (2–6 keV) 1.9×10^{38} erg s^{-1}.
Optics: $V = 12.02$ B5 I type, $B − V = 0.24$, $U − B = −0.57$, $M_v = −7.6$, $A_v = 0.96$.
(Cooke *et al.*, 1978; Einstein, 1977; Forman *et al.*, 1978; Johnston *et al.*, 1978; Markert and
Clark, 1975a; Markert *et al.*, 1978; Schnopper and Delvaille, 1977b).

05418+608   Forman *et al.* (1978).

05438−316   Cluster of galaxies (?).
If it is SC 0549−326, then luminosity (2–6 keV) 4.8×10^{44} erg s^{-1} (radio source PKS 0548
−322). If it is SC 0550−316, then luminosity (2–6) keV 3×10^{44} erg s^{-1} (radio source PKS
0548−317).
(Cooke *et al.*, 1978; Forman *et al.*, 1978; Lugger, 1978; Markert *et al.*, 1978; Murray and Ulmer,
1976).

05441−665   X-rays: luminosity (2–11 keV) 1.5×10^{37} erg s^{-1}.
Optics: blue star, $B \sim 16$, $M_B \sim −3$ (identification is not safe).
(Bradt *et al.*, 1978; Johnston *et al.*, 1979a).

05464−882   Forman *et al.* (1978).

05480+290   Forman *et al.* (1978).

05488−332   X-rays: spectrum $T = 4 \times 10^6$ K, $N_H \sim 10^{20}$ cm^{-2}, luminosity (2–6 keV) 6×10^{44} erg s^{-1}.
Optics: BL Lac-type object, $B \approx 16.1$, $Z = 0.042$.
(Riegler *et al.*, 1978a; Schwartz *et al.*, 1978).

05497−074   Galaxy NGC 2110.
X-rays: power spectrum, index $−0.1 \pm 0.3$, luminosity (2–11 keV) 1.2×10^{43} erg s^{-1}.
Optics: elliptical galaxy NGC 2110, $z = 0.0071 \pm 0.0003$, distance 43 Mpc.
Radio: fluxes 0.24 J (1.48 GHz) and 0.13 J (4.88 GHz).
(Bradt *et al.*, 1979; Griffiths *et al.*, 1979; McClintock *et al.*, 1979; Marshall *et al.*, 1979).

05512+466   Seyfert galaxy NGG 8-II-II (?).
Possibly it is X-ray source 06002+465 (?).
X-rays: luminosity (2–6 keV) $\sim 1.2 \times 10^{44}$ erg s^{-1}.
Optics: $V = 14.7$.
(Cooke *et al.*, 1978; Forman *et al.*, 1978; Griffiths *et al.*, 1979; Miller, 1979).

05538−486   Forman *et al.* (1978).

Table II (continued)

05570−381	Forman *et al.* (1978).
05598−571	Forman *et al.* (1978).
06002+465	Possibly, it is X-ray source 05512+466 (?).

06002+465 (Markert *et al.*, 1975b, 1976a, 1978; Murray and Ulmer, 1976).

06084−491 Forman *et al.* (1978).

06088+497 Cooke *et al.* (1978).

06137+224 SNR IC 443.
X-rays: spectrum $kT = 1.4 \pm 0.3$ keV, $N_H = 1.8 \times 10^{21}$ cm^{-2}, luminosity (0.5–2 keV) $\sim 10^{35}$ erg s^{-1}, age ~ 3400 yr.
Optics: diam 20 pc, $E_\odot \approx 4 \times 10^{50}$ erg, distance 1.5–2.5 kpc.
Radio: spectral index $\alpha = -0.4$, flux 180 J (1 GHz), enveloped structure.
(Culhane, 1977; Gorenstein and Tucker, 1975; Lozinskaya, 1975; Markert *et al.*, 1976; Markert *et al.*, 1978; Parkes *et al.*, 1977; Winkler *et al.*, 1974; Winkler, 1978; Woltjer, 1972).

06143+091 X-rays: in September 1975 flare of duration 40 s, $kT = 1.1$–2.5 keV. Stable source, $kT \simeq 3$–4 keV, eclipses are not observed, variability of Sco X-1 type.
Optics: variable star $V = 18.75$, emission of C III and N III, distance 4–8 kpc.
(Davidsen *et al.*, 1974; Forman *et al.*, 1978; Jones, 1976; Markert *et al.*, 1978; Mason *et al.*, 1976c; Murdin *et al.*, 1974; Parsignault and Grindlay, 1978; Swank *et al.*, 1978; Willmore *et al.*, 1974).

06148+153 Forman *et al.* (1978).

06150+093 Possibly, it is X-ray source 06143+091 (?).
X-rays: flare of duration ~ 100 min, soft spectrum.
(Lewin, 1976a).

06201−003 X-rays: transient source was discovered 3.8.75.
(Ariel). Spectrum softened from $kT \sim 30$ keV (4–6.8.75) to $kT = 1.3$ keV (10–16.8.75). X-ray orbital period 7.8 $\pm$ 0.7 days observed (SAS-3). $N_H = 5 \times 10^{22}$ cm^{-2} (in maximum).
Optics: star of $V = 11$ (12.8.75). Palomar plates give $B = 20.5$. Star flared in February 1917. Possible spectral type M5 IV, $M_v = 5.2$, distance ~ 1 kpc. Possibly, a period of 3.92 $\pm$ 0.02 days observed. Emission lines Hα, He I 4471 and absorption lines Hα and Hβ observed. Flux (0.6–0.9 keV) 4.14 sq cm $^{-2}$ s^{-1} (Saljut–75/9/10, 4).
Radio: flare in 75/8, maximum 0.3 f.u. (1400 GHz).
(Bieging and Downes, 1975; Boley *et al.*, 1976; Bortle, 1976; Bradt and Matilsky, 1976; Ciatti *et al.*, 1977; Citterio *et al.*, 1976; Chi-Chao *et al.*, 1976b; Eachus *et al.*, 1976; Cominsky *et al.*, 1978; Elvis *et al.*, 1975; French, 1975; Gull *et al.*, 1976; Kaluzienski *et al.*, 1976b, c; Kestenbaum *et al.*, 1976; Kirshner, 1975; Kleinmann, 1976; Kleinmann *et al.*, 1976; Kleinmann *et al.*, 1976b, c; Kurt *et al.*, 1976; Lyutyi, 1976; Markert *et al.*, 1977; Martinov, 1976; Matilsky, 1976; Matilsky *et al.*, 1976; Owen *et al.*, 1976; Peterson, 1975; Ricketts *et al.*, 1975b; Searle, 1975; Ward *et al.*, 1975; Wyckoff and Wehinger, 1975).

06216+117 Forman *et al.* (1978).

06225−529 X-rays: flux (0.4–3 keV) (3.1 $\pm$ 0.7) $\times$ 10^{-11} erg cm^{-2} s^{-1}.
(Riegler *et al.*, 1978).

06261−541 Cluster of galaxies SC 0627−544, SG 0630−551 (?).
X-rays: spectrum $kT = 6.3$ (+3.7; −1.9) keV, luminosity (2–10 keV) 8.4 $\times$ 10^{44} erg s^{-1}.
(Cooke *et al.*, 1978; Forman *et al.*, 1978; Lugger, 1978; Markert *et al.*, 1978; McHardy, 1978; Murray and Ulmer, 1976; Mushotzky *et al.*, 1978).

06270+216 Star.
Optics: flares, $\Delta m = 3.2$. In quiet state $V = 14.9$.
(Watson *et al.*, 1978).

06272+675 Forman *et al.* (1978).

06273−381 Forman *et al.* (1978).

06288−284 Forman *et al.* (1978).

06300+024 Forman *et al.* (1978).

06350−033 Forman *et al.* (1978).

06384+472 Seyfert galaxy MKN 6 (?).
(Forman *et al.*, 1978; Ricketts, 1978).

Table II (continued)

06429 − 166	α CMa (Sirius). X-rays: Intensity 10^{-11} erg cm^{-2} s^{-1} ($\sim$0.2 keV), luminosity 10^{28} erg s^{-1}, $T \sim 2.3 \times 10^5$ K. Optics: $V = -1^m46$. Spectrum A IV, binary system with white dwarf. $P_{orb.} = 50$ yr. (Mewe *et al.*, 1975; Lampton *et al.*, 1979).
06430 + 534	Galaxy (possible Seyfert). Anon 0636 + 53 (?). X-ray: luminosity 10^{44} erg s^{-1}. (Marshall *et al.*, 1979).
06560 − 071	Transient source (?). X-rays: crab-like spectrum, $kT = 5 \pm 2$ keV. Optics: Be star (?), $V = 12.35$, $B - V = 0.87$; $U - B = -0.116$, emission Hα, Hβ. (Carpenter *et al.*, 1975; Clark, 1975b; Cominsky *et al.*, 1978; Kaluzienski *et al.*, 1976b; Markert *et al.*, 1977; Pakull, 1978).
06560 − 031	Forman *et al.* (1978).
06572 − 114	Watson *et al.* (1978).
06576 − 351	Markert *et al.* (1978).
07002 − 563	Possibly, there are 2 sources on this position (?). (Marshall *et al.*, 1979).
07056 + 186	X-rays: spectrum $kT = 6 \pm 3$ keV, $N_H < 4 \times 10^{22}$ cm^{-2}. (Lamb and Worral, 1979).
07080 − 357	Cluster of galaxies SG 0659 − 349 (?). Radio source PKS 0657 − 348 (?). X-rays: luminosity (2–10 keV) $\sim 1.1 \times 10^{46}$ erg s^{-1}. (Cooke *et al.*, 1978; Giacconi *et al.*, 1974; Lugger, 1978).
07083 − 168	Forman *et al.* (1978).
07084 − 492	Forman *et al.* (1978).
07090 − 221	Watson *et al.* (1978).
07107 − 456	Seyfert galaxy MKN 376.(?). (Cooke *et al.*, 1978; Forman *et al.*, 1978).
07113 − 384	Forman *et al.* (1978).
07124 − 113	Marshall *et al.* (1979).
07183 − 546	Cooke *et al.* (1978), Forman *et al.* (1978).
07202 + 558	Cluster of galaxies Abell 576 (?). X-rays: luminosity (2–10 keV) $\approx 3.7 \times 10^{44}$ erg s^{-1}. (Forman *et al.*, 1978, 1978a; Ricketts, 1978).
07267 − 260	Watson *et al.* (1978), Forman *et al.* (1978).
07284 + 060	Flare star SVCMi. Optics: flares $\Delta m = 3.3$, in stable state $V = 16.3$. (Watson *et al.*, 1978).
07290 − 379	Forman *et al.* (1978).
07333 − 186	Forman *et al.* (1978).
07360 − 500	X-rays: duration of burst ~ 15 s. (Doty, 1976).
07373 − 108	Forman *et al.* (1978).
07380 + 498	Seyfert galaxy MKN 79 (??). (Cooke *et al.*, 1978).
07390 − 199	Forman *et al.* (1978), Watson *et al.* (1978).
07394 + 036	Flare star YZ CMi. X-rays: duration of burst $\sim$6 min. Energy of burst $\sim 2 \times 10^{32}$ erg (0.2–7 keV). Optics: UV Ceti-type star, spectrum dM 4.5e, distance 6.06 pc. Optical flare 19.10.74 not observed. (Heise *et al.*, 1975).
07401 + 290	X-rays: luminosity (0.2–2.8 keV) $\approx 1.3 \times 10^{32}$ erg s^{-1} (?). Optics: star HD 62044 (?), $V = 4.26$, spectrum K 1 III, orbital period of 19 605 days, distance 150 pc. (Woltjer *et al.*, 1978a).

Table II (continued)

07429 − 286 Forman *et al.* (1978), Markert *et al.* (1977).

07517 + 222 Marshall *et al.* (1979).

07521 + 221 Flare of U Gem.
X-rays: on 19.10.77. upper limit of intensity 10^{-11} erg cm^{-2} s^{-1} was measured. On 20.10.77 intensity was $\sim 10^{-11}$ erg cm^{-2} s^{-1} (0.15–0.30 keV). Spectrum $kT \sim 0.04$ keV.
Optics: in quiet state $V = 14.5$, $M_v = 8.8$ (flare), $\Delta V = 5.6$. Flare was observed 19–20.10.77.
(Garmir *et al.*, 1977; Mason *et al.*, 1978a, b; Swank *et al.*, 1978b).

07574 − 484 X-rays: 1800 K $< T_{\text{eff}} < 27\,000$ K.
Optical candidates: (i) eclipsing binary system *V* Pup, strong emission, $V = 4.3$, $P_{\text{orb.}} = 1\,3295$ days, (ii) helium variable star HD 64740 (spectrum *B* IV), mass $> 2M_\odot$, $V \sin i = 150$ km s^{-1}.
(Bahcall *et al.*, 1975; Cooke *et al.*, 1978; Forman *et al.*, 1978; Groote *et al.*, 1978; Markert *et al.*, 1978; Pedersen, 1976; Seward *et al.*, 1976b).

07578 − 264 Markert *et al.* (1978).

08048 − 530 Markert *et al.* (1978).

08081 − 352 Possibly, it is Nova 1942 CP Pup (?).
X-rays: intensity 8.9×10^{-12} erg cm^{-2} s^{-1} $\approx$ (0.15–0.5 keV), in the range of 0.2–2 keV, the upper limit of intensity is 4.4×10^{-12} erg cm^{-2} s^{-1}, soft spectrum.
(Jensen, 1979).

08134 − 385 Forman *et al.* (1978).

08142 − 567 Forman *et al.* (1978), Seward *et al.* (1976a).

08152 − 075 Cluster of galaxies Abell 644 (?).
(Cooke *et al.*, 1978).

08215 − 427 SNR PuP A.
X-rays: spectrum $kT = 0.5$ keV, $N_{\text{H}} = 3.5 \times 10^{21}$ cm^{-2}, luminosity (0.5–10 keV) $\sim 2.5 \times 10^{36}$ erg s^{-1}, $E_0 \approx 3 \times 10^{50}$ erg.
Optics: diam 10–20 pc, distance 1.2–2.2 kpc.
Radio: age 4–5 thousand yrs.
flux 145 f.u. (1 GH$_z$).
(Catura and Acton, 1976a; Culhane, 1977; Forman *et al.*, 1978; Gorenstein *et al.*, 1974; Markert *et al.*, 1978; Seward *et al.*, 1976; Smith, 1978).

08336 − 450 SNR Vela X.
X-ray: spectrum of remnant $kT = 0.3$ keV, luminosity (0.5–2 keV) $\sim 3 \times 10^{35}$ erg s^{-1}, Pulsar $P = 0.0892$ s. 23.7.76 flare was observed ≈ 12 h, $kT = 0.2$–0.3 keV.
Optics: diam 20–40 pc, distance 0.5–1 kpc.
Radio: age 1.1×10^4 yr, flux 1800 f.u. (1 GHz). Possibly, there is a connection with gamma-source CG 263–2.
(Becker *et al.*, 1978; Culhane, 1977; Forman *et al.*, 1978; Gorenstein *et al.*, 1974; Hunt *et al.*, 1979; Markert *et al.*, 1978; Marshall *et al.*, 1977; Mason and Culhane, 1978; Moore and Garmire, 1976; Pravdo *et al.*, 1976; Seward *et al.*, 1976a; Smith, 1976; Thompson *et al.*, 1977; Winkler, 1978).

08356 − 483 Possibly, it is 08514 − 469.
(Forman *et al.*, 1978; Seward *et al.*, 1976a; Smith, 1978).

08362 − 426 Transient source.
X-ray: spectrum $kT = 3$–7 keV, $E_a \leqslant 2.6$ keV, $N_{\text{H}} < 3.5 \times 10^{22}$ cm^{-2}.
(Cominsky *et al.*, 1978; Forman *et al.*, 1978; Johns, 1976; Markert *et al.*, 1975b, 1977, 1978; Smith, 1978; Villa *et al.*, 1976).

08429 − 349 Forman *et al.* (1978).

08449 − 531 Marshall *et al.* (1979).

08450 − 296 Forman *et al.* (1978).

08514 − 469 Possibly, it is 08356 − 483.
(Watson *et al.*, 1978).

08542 − 445 Forman *et al.* (1978), Smith (1978).

08590 + 509 Cooke *et al.* (1978).

09002 − 403 Binary system HD 77531.

Table II (continued)

	X-rays: eclipse, $P_{orb.}$ = 3.9643 days, pulsar P = 282. 9017 s, excentrisity of the system 0.079 $\pm$ 0.014, K_x = 275.8 km s^{-1}, $a \sin i$ = 3.4 $\times$ 10^{12} cm, spectrum kT > 15 keV.

X-rays: eclipse, $P_{orb.}$ = 3.9643 days, pulsar P = 282. 9017 s, excentrisity of the system 0.079 $\pm$ 0.014, K_x = 275.8 km s^{-1}, $a \sin i$ = 3.4 $\times$ 10^{12} cm, spectrum kT > 15 keV.
Optics: V = 6.9, ΔV = 0.07, M_B = $-$8.4, spectrum B 0.5 Ia. Distance 1.3 $\pm$ 0.2 kpc, $M_{opt.}$ = 21.2 $\pm$ 2.4 $M_\odot$, M_x = 1.6 $\pm$ 0.2 $M_\odot$.
(Avni and Bahcall, 1975; Charles *et al.*, 1978; Class, 1979; Gorenstein and Tucker, 1972; Hill *et al.*, 1972; Hutchings, 1978; Johns, 1976; Markert *et al.*, 1978; McClintock and Rappaport, 1976; McClintock *et al.*, 1976; Rappaport and McClintock, 1976; Seward, 1970; Steiner, 1977; van Paradijs *et al.*, 1976; Vidal, 1976).

09020 + 573 Duration of burst is less than 128 s.
(Evans *et al.*, 1970).

09062 − 095 Cluster of galaxies Abell 754 (?).
(Cooke *et al.*, 1978; Forman *et al.*, 1978; McHardy, 1978; Markert *et al.*, 1978).

09084 − 669 Forman *et al.* (1978).

09132 − 461 Forman *et al.* (1979).

09177 + 634 Bahcall (1976b), Cooke *et al.* (1978), Giacconi *et al.* (1974), Markert *et al.* (1976a, 1978).

09191 − 549 Bradt (1978), Forman *et al.* (1978), Markert *et al.* (1978), Seward *et al.* (1976a).

09207 − 628 Watson *et al.* (1978).

09214 − 630 X-rays: luminosity (2–10 keV) $\sim$ 6 $\times$ 10^{35} (D/10)^2 erg s^{-1}.
Optics: star V = 17 strong emission Hα, He II 4686 and weak emission Hβ. $L_x/L_{opt.}$ $\sim$ 10.
(Li *et al.*, 1978; Marshall *et al.*, 1978).

09233 − 314 Cooke *et al.* (1978), Forman *et al.* (1978), Schwartz *et al.* (1978).

09433 − 140 Possibly, interacting system NGC 2993/2 (?).
(Cooke *et al.*, 1978; Forman *et al.*, 1978; Griffiths *et al.*, 1979).

09436 + 712 Markert *et al.* (1978).

09459 − 306 Cluster of galaxies SC 0948 − 327.
X-rays: luminosity (2–10 keV) $\sim$ 3.7 $\times$ 10^{44} erg s^{-1}.
Radio: source PKS 0947 − 824.
(Cooke *et al.*, 1978; Forman *et al.*, 1978; Griffiths *et al.*, 1979; Lugger, 1978; Markert *et al.*, 1978; Pineda and Schnopper, 1978).

09483 + 121 Star X Leo, variability ΔV = 3.4
(Watson *et al.*, 1978).

09544 + 700 Possibly, galaxy M82 (?).
X-rays: luminosity (2–6 keV) $\sim$ 10^{40} erg s^{-1}.
(Cooke *et al.*, 1978; Forman *et al.*, 1978; Griffiths *et al.*, 1979).

09555 − 284 Forman *et al.* (1978).

09587 − 359 Cluster of galaxies SG 0959 − 359.
X-rays: luminosity (2–6 keV) $\sim$ 1.5 $\times$ 10^{45} erg s^{-1}.
Radio: source OL-302.
(Lugger, 1978).

10088 + 138 Cluster of galaxies Abell 999.
(Ricketts, 1978).

10144 − 579 Seward *et al.* (1976a).

10151 − 254 Cluster of galaxies Abell 955, 956.
(Forman *et al.*, 1978; Ricketts, 1978).

10184 + 498 Ricketts (1978).

10220 − 408 Forman *et al.* (1978).

10224 − 554 Possibly, XRS 10346 − 565 (?).
(Giacconi *et al.*, 1974; Markert *et al.*, 1978; Villa *et al.*, 1976).

10270 − 590 X-ray: luminosity $\sim$ 10^{31} erg s^{-1} (0.1–0.3 keV).
Optics: flare star dM Oe, distance 20.8 pc.
(Lategan, 1978).

10284 + 512 Galaxy MKN 142 (?).
(Marshall *et al.*, 1979).

10335 − 270 Cluster of galaxies Abell 1060.

Table II (continued)

X-ray: spectrum $kT = 3.1$ keV, $N_H < 6.3 \times 10^{21}$ cm^{-2}, luminosity (2–10 keV) $\sim 2 \times 10^{43}$ erg s^{-1}.
(Cooke *et al.*, 1978; Forman *et al.*, 1978; Henry and Tucker, 1979; Markert *et al.*, 1978; McHardy, 1978; Mushotzky *et al.*, 1978).

$10346 - 565$ Possibly, it is $10224 - 557$.
(Forman *et al.*, 1976a, 1977; Seward *et al.*, 1976a; Watson *et al.*, 1978).

$10413 - 079$ Cooke *et al.* (1978).

$10416 - 218$ Forman *et al.* (1978).

$10440 - 594$ X-ray: spectrum $kT = 7.3$ keV, $E_a = 0.09$ keV, luminosity (2–6 keV) $\sim 1.5 \times 10^{36}$ erg s^{-1}.
Optics: η Car in 1843 yr: $V = -1$, in 1971: $V = 7$. Distance 1.6 kpc.
(Becker *et al.*, 1976; Forman *et al.*, 1978; Hill *et al.*, 1972; Seward *et al.*, 1976a, b).

$10450 - 593$ Supernova remnant G 287.8–0.5.
X-ray: luminosity (1–10 keV) $\sim (0.3–1) \times 10^{35}$ erg s^{-1}, spectrum $kT = 6$ keV, age ~ 500 yr.
Radio: distance 2.5 kpc, diam 35 pc.
(Culhane, 1977; Winkler, 1978).

$10520 + 560$ X-ray: intensity (0.2–0.4 keV) $\sim 2 \times 10^{-10}$ erg s^{-1} cm^{-2}, luminosity (0.2–0.4 keV) $\sim 6.5 \times 10^{30}$ erg s^{-1}.
Optics: dwarf M1, distance 15.9 pc.
(Lategan, 1978).

$10528 + 606$ X-ray: luminosity (2–6 keV) $\sim 10^{32}$ erg s^{-1}.
Optics: star SAO 015338; spectrum K 1 IV, type RS CVn strong emission Hα, $P_{orb.} = 7.5$ days, $k \sim 30$ km s^{-1}, $V = 8.8$, $M_v = 3.2$, distance 160 pc.
(Cooke *et al.*, 1978; Crampton, 1978; Giacconi *et al.*, 1974; Liller, 1978; Schwartz *et al.*, 1978, 1979).

$10586 - 226$ Cluster of galaxies Abell 1146 (?).
(Cooke *et al.*, 1978; Forman *et al.*, 1978; McHardy, 1978).

$10598 + 384$ Possibly, it is source $11016 + 384$ (?).
Galaxy MKN 421 (?).
X-ray: spectrum $kT = 0.15–0.35$ keV, $N_H = (1–4) \times 10^{20}$ cm^{-2}.
(Woltjer and Mason, 1979).

$11015 + 450$ X-ray: intensity (0.1–0.4 keV) $\sim 2.8 \times 10^{-11}$ erg cm^{-2} s^{-1}, spectrum $kT = 12–30$ keV, $N_H < 2 \times 10^{20}$ cm^{-2}.
Optics: star AN UMA.
(Hearn and Marshall, 1979).

$11016 + 384$ Galaxy MKN 421 (?), type BL Lac.
X-ray: burst $\sim$ days, spectral index -2.1, $N_H = 3 \times 10^{20}$ cm^{-2}, luminosity (2–6 keV) $\sim 3 \times 10^{45}$ erg s^{-1}.
Optics: $B = 16.3$, distance 180 mpc.
(Cooke, 1976; Cooke *et al.*, 1978; Hearn and Marshall, 1978; Hearn *et al.*, 1979; Lawrence *et al.*, 1977; Marshall and Jernigan, 1978; Schwartz *et al.*, 1978).

$11095 + 597$ Markert *et al.* (1978).

$11102 - 580$ Forman *et al.* (1978).

$11110 - 603$ X-ray: luminosity (0.37–1.9 keV) in the burst 9×10^{30} erg s^{-1}.
Optics: flare star dM0e, distance 11.1 pc.
(Lategan, 1978).

$11189 - 615$ Transient source.
X-ray: burst ~ 7 days. Pulsar $P = 6.755$ min. Spectrum $kT > 15$ keV, $N_H = 4 \times 10^{22}$ cm^{-2}.
(Carpenter *et al.*, 1977; Cominsky *et al.*, 1978; Eyles *et al.*, 1975a; Ives *et al.*, 1975; Johns, 1976; Markert *et al.*, 1977; Willmore, 1977).

$11190 - 603$ Star V 779 Cen.
X-ray: eclipsing-binary system, $P_{orb.} = 2.086909$ days, excentricity $e = 0.0008 \pm 0.0001$, duration of the eclipse 0.488 ± 0.012 days. Pulsar $P = 4.83704$ s, spectrum $kT = 20$ keV, $N_H = 8.8 \times 10^{21}$ cm^{-2}, $E_a \doteq 2.0$ keV, luminosity (2–6 keV) 6×10^{37} erg s^{-1}.

Table II (continued)

	Optics: $V = 13.4$, $B - V = 1.05$; $M_v = -5.9$, spectrum O 6.5 eqv. V-III, sin $i = 0.985 \pm 0.02$, $K = 24 \pm 6$ km s^{-1}, $a = 1.2 \times 10^{12}$ cm, masses of components 17 ± 2 $M_\odot$ and 1.0 ± 0.3 $M_\odot$ (X-ray source). $A_v = 4.2$.
	(Avni and Bahcall, 1976; Bunner and Sanders, 1976; Coe *et al.*, 1976a; Forman *et al.*, 1978; Henry and Schreier, 1977; Johns, 1976; Krzeminski, 1974; Lamb *et al.*, 1976; Mauder, 1976a; Osmer *et al.*, 1975; Ryter, 1976; Schreier and Fabbiano, 1976; Seward *et al.*, 1976a; Swank *et al.*, 1976d).
11196 − 778	Forman *et al.* (1978).
11206 − 431	Forman *et al.* (1978).
11222 − 590	SNR MSH II-54 (?), GX 292.0 ± 1.8.
	(Share *et al.*, 1978).
11304 − 146	Cluster of galaxies Abell 1285 (?).
	X-ray: luminosity (2–6 keV) $\sim 3 \times 10^{45}$ erg s^{-1}.
	(Forman *et al.*, 1978; Ricketts, 1978; Schwartz, 1978).
11353 + 525	X-ray: luminosity (0.2–2.8 keV) $\sim 9.6 \times 10^{32}$ erg s^{-1}.
	Optics: RW UMa, star of RS CVn type, $V = 10.3$–11.9, $P = 7.328\,23$ days, spectrum dF9 + + dG9, distance 150 pc.
	(Walter *et al.*, 1978a).
11357 − 373	Seyfert galaxy NGC 3783 (?).
	(Cooke *et al.*, 1978; Forman *et al.*, 1978).
11373 − 651	Forman *et al.* (1978), Markert *et al.* (1978), Seward *et al.* (1976a).
11432 − 185	Ricketts (1978).
11440 + 197	Cluster of galaxies Abell 1367 (?).
	X-ray: spectrum $kT = 2.8$ keV, $N_H = 4.2 \times 10^{22}$ cm^{-2}, luminosity (2–10 keV) $\sim 1.3 \times 10^{44}$ erg s^{-1}.
	(Cooke *et al.*, 1978; Forman *et al.*, 1978, 1978a; Henry and Tucker, 1979; Markert *et al.*, 1978; McHardy, 1978; Mushotzky *et al.*, 1978; Schwartz, 1978).
11440 + 840	Forman *et al.* (1978).
11448 − 748	Markert *et al.* (1978).
	X-rays: spectrum $kT = 7$–15 keV, $E_a = 1.5$–3.5 keV, $N_H = 1.2 \times 10^2$ cm^{-2}, pulsar $P = = 297.24$ s, luminosity (2–6 keV) $\sim 6 \times 10^{36}$ erg s^{-1}.
	Optics: star Hen 715, $V = 8.9$, spectrum BI Vne, distance 1.5 kpc, $A_v \sim 1.6$, variable emission Hα.
	(Bradt *et al.*, 1978; Forman *et al.*, 1978; Class, 1979; Hill *et al.*, 1972; Johns, 1976; Jernigan *et al.*, 1978; Markert *et al.*, 1978; Seward *et al.*, 1976a; Villa *et al.*, 1976; White, 1977).
11474 − 124	Cluster of galaxies Abell 1391 (?).
	(Forman *et al.*, 1978; Ricketts, 1978).
11500 + 720	Cluster of galaxies Abell 1254 (?).
	X-rays: spectrum $kT = 18.4$ keV, $N_H \approx 5.4 \times 10^{21}$ cm^{-2}.
	(Cooke *et al.*, 1978; Mushotzky *et al.*, 1978).
11508 + 745	Duration of burst 80 s.
	(Wheaton *et al.*, 1973).
11536 − 115	Forman *et al.* (1978).
11539 − 402	Forman *et al.* (1978).
11576 − 071	Ricketts (1978).
12036 − 061	Forman *et al.* (1978).
12070 − 521	Supernova remnant PRS 1209 − 52.
	X-rays: extended, in the range of 0.2–1 keV, flux 1.4×10^{-10} erg cm^{-2} s^{-1}, spectrum $kT = 0.1$ keV, $N_H = 3 \times 10^{21}$ cm^{-2}, luminosity (0.2–1 keV) $\sim 8 \times 10^{35}$ erg s^{-1}.
	Radio: flux 55 Jy at 1 GHz, $\alpha = -0.45$, $D = 1.1$–2 kpc, $R = 13$–24 pc.
	(Tuchy *et al.*, 1978, 1979).
12078 + 397	Seyfert galaxy NGC 4151.
	X-ray: slow bursts ~ 1400 days, rapid bursts ~ 10 days. Spectral index -1.42, $N_H \simeq 7.5 \times 10^{22}$ cm^{-2}, luminosity 3.6×10^{42} erg s^{-1}.

Table II (continued)

	Optics: distance 20 mpc.
	(Auriemma *et al.*, 1978; Cooke *et al.*, 1978; Forman *et al.*, 1978; Griffiths *et al.*, 1979; Lawrence *et al.*, 1977; Markert *et al.*, 1978; Stanbert *et al.*, 1978).
12096−452	Forman *et al.* (1978).
12103−646	Forman *et al.* (1978), Seward *et al.* (1976a), Markert *et al.* (1978).
12150+440	X-ray: soft X-ray flare in the maximum of luminosity (0.2–0.4 keV) $\sim 1.8 \times 10^{30}$ erg s^{-1} (at the distance 12 pc).
	Optics: dwarf dK4, distance 12 pc, bursting star.
	(Lategan, 1978).
12156−594	Seward *et al.* (1976a).
12170−672	Flare with duration less than 128 s.
	(Evans *et al.*, 1976).
12190+305	Object of BL Lac type (?).
	X-ray: spectral index -0.5 ± 0.5.
	Optics: $V = 16.4$, variability $\Delta m = 1.1$, $U - B = -0.50$; $B - V = 0.65$; $U - V = 0.15$.
	Radio: on 1400 mHz flux 0.054 Jy.
	(Cooke *et al.*, 1978; Schwartz *et al.*, 1978; Wilson *et al.*, 1979).
12200+260	X-ray: flare in the range of 0.1–0.4 keV, flux $\sim 7 \times 10^{-9}$ erg cm^{-2} s^{-1}, luminosity 3.4×10^{32} erg s^{-1}.
	Optics: flare star of dM2 type, distance ~ 19.2 pc.
	(Lategan, 1978).
12214−081	Cooke *et al.* (1978), Forman *et al.* (1978).
12238−624	X-ray: pulsar $P = 695.7$ s, spectrum is hard $kT > 15$ keV, $N_H = (0.5–1.7) \times 10^{24}$ cm^{-2}, luminosity 4×10^{36} erg s^{-1} (distance 2 kpc).
	Optics: star WRA 977, $V = 10.84$, spectrum B 1.5 Ia, binary $P_{\text{orb.}} = 35$ days, activity $\sim$ days and hours, $M_v = -6$, distance 2 kpc, $a \sin i \sim 9 \times 10^{12}$ cm, $e = 0.44$, pulsar $P = 699.4$ s.
	(Bord *et al.*, 1976; Bradt *et al.*, 1978; Coe *et al.*, 1976a; Cooke and Pounds, 1971; Forman *et al.*, 1978; Glass, 1979; Johns, 1976; Kelley *et al.*, 1979; Kelley and Bradt, 1978; Lewin *et al.*, 1971; Markert *et al.*, 1978; Mauder, 1976; Pakull, 1978; Ricker *et al.*, 1976a; Seward *et al.*, 1976a; Swank *et al.*, 1976a; van Gederen 1977; Vidal, 1973; White *et al.*, 1975, 1976b; White *et al.*, 1978; Willmore *et al.*, 1979).
12260+024	Quasar 3C 273.
	X-ray: spectral index -0.4 ± 0.2, luminosity (2–40 keV) 1.4×10^{46} erg s^{-1}.
	Optics: $V = 12.80$, distance ~ 950 mpc, $A_v = 0.20$.
	Radio: on 1.0 GHz flux 40 f.u. Gamma-region: on 100 meV flux 8×10^{-6} f.u.
	(Bradt *et al.*, 1979; Bradt *et al.*, 1979; Cooke *et al.*, 1978; Forman *et al.*, 1978; Markert *et al.*, 1978; Primini *et al.*, 1978; Swanenburg *et al.*, 1978).
12270+198	X-ray: burst with duration less than 128 s.
	(Evans *et al.*, 1976).
12287+126	Cluster of galaxies Virgo.
	X-ray: spectrum $kT = 2.6$ keV, $N_H \approx 2 \times 10^{21}$ cm^{-2}, luminosity (2–10 keV) $\sim 3 \times 10^{43}$ erg s^{-1}.
	(Cooke *et al.*, 1978; Davison, 1978; Forman *et al.*, 1978; Forman *et al.*, 1978a; Friedman *et al.*, 1967; Henry and Tucker, 1979; Li *et al.*, 1978; Markert *et al.*, 1978; McHardy, 1978; Mushotzky *et al.*, 1978).
12329+071	Cooke *et al.* (1978), Forman *et al.* (1978), Markert *et al.* (1978),
12391−599	X-ray: pulsar $P = 191$ s (?), spectrum hard, index -0.8.
	Optics: IR – object, $K = 9.63$, $\Delta K = 1.92$.
	(Bradt, 1978; Bradt *et al.*, 1978; Carpenter *et al.*, 1977; Glass, 1979).
12403−056	Cluster Abell 1588 (?).
	(Forman *et al.*, 1978; Marshall *et al.*, 1979; Ricketts, 1978).
12446−603	Source was observed only during 0.5 days.
	(Carpenter *et al.*, 1979).
12462−410	Cluster of galaxies in Cen (NGC 4696).

Table II (continued)

	X-ray: spectrum $kT = 3.5$ keV, luminosity (2–6 keV) $\sim 6 \times 10^{43}$ erg s^{-1}. Radio: source PKS 1245–41. (Cooke *et al.*, 1978; Forman *et al.*, 1978; Henry and Tucker, 1979; Markert *et al.*, 1978; McHardy, 1978; Mushotzky *et al.*, 1978).
12466–588	Possibly, very rapid transient source. (Carpenter *et al.*, 1977; Kaluzienski, 1977; Seward *et al.*, 1976a).
12492–289	X-ray: spectrum $kT = 4.5$ keV, luminosity 10^{32} erg s^{-1} $\sim$ (2–10 keV). Optics: EX Hydrae, binary system $P_{orb.} = 99$ min, bursts with the interval ~ 465 days, duration ~ 4 days. $V = 13.7$–11.5, distance 100 pc, $B - V = -0.7$, $U - V = -1$. (Bradt *et al.*, 1978; Cooke *et al.*, 1978; Cordova and Riegler, 1978; Forman *et al.*, 1978; Johnson, 1978; Lugger, 1978; Markert *et al.*, 1978; McHardy, 1978; Mushotzky *et al.*, 1978; Schwartz *et al.*, 1979).
12500–667	Seward *et al.* (1976a).
12539–002	Forman *et al.* (1978).
12543–690	X-rays: spectrum $kT = 3$–7 keV, $E_a \leqslant 2.6$ keV. Optics: star $V = 19.09$, $B = 19.24$ (?), distance ~ 10 kpc. (Coe *et al.*, 1976a; Forman *et al.*, 1978; Griffiths *et al.*, 1978; Johns, 1976; Markert *et al.*, 1978; Seward *et al.*, 1976a).
12544–169	Possibly, is XRS 12566–171 (?). Cluster of galaxies Abell 1644 (?). (Ricketts, 1978).
12566–171	Possibly it is XRS 12544–169 (?). Cluster of galaxies Abell 1644 (?). (Marshall *et al.*, 1979).
12574+283	Cluster of galaxies Abell 1656. X-rays: $kT = 5.6$ keV, $N_H = 3.1 \times 10^{21}$ cm^{-2}, extended source, diam 1050 kpc, luminosity (2–10 keV) $\sim 8 \times 10^{44}$ erg s^{-1}. (Cooke *et al.*, 1978; Forman *et al.*, 1978; Forman *et al.*, 1978; Henry and Tucker, 1979; Markert *et al.*, 1978; McHardy, 1978; Mushotzky *et al.*, 1978).
12582–613	X-rays: pulsar $P = 272.2$ s, spectrum $kT = 7$–15 keV, $E_a = 2.6$ keV, luminosity (2–6 keV) $\sim 2 \times 10^{36}$ erg s^{-1}. Optics: star $V = 14.7$, $B - V = 1.8$, $A_v = 6.3$, spectrum B0-B5V distance ~ 2 kpc. (Bradt *et al.*, 1977a; Bradt *et al.*, 1978; Coe *et al.*, 1976a; Forman *et al.*, 1978; Glass, 1979; Johns, 1976; Markert *et al.*, 1978; McClintock *et al.*, 1977a; Mason *et al.*, 1978; Seward *et al.*, 1976a; Villa *et al.*, 1976).
13007–488	Forman *et al.* (1978).
13020–775	Forman *et al.* (1978).
13061–012	Cooke *et al.* (1978).
13083+382	X-ray: in the range 0.2–2.8 keV, intensity 0.01 keV cm^{-2} s^{-1}, luminosity $\sim 10^{32}$ erg s^{-1}. Optics: star HD 114519, type RS CVn, period $P = 4.797\,891$ days. $V = 8.4$, spectrum F4 IV – V + K0 IV. (Woltjer *et al.*, 1978b).
13088+864	Forman *et al.* (1978).
13108+371	Emission galaxy NGC 5005, 5033 (?). (Marshall *et al.*, 1979).
13140+292	X-ray: intensity $\sim 3 \times 10^{-10}$ erg cm^{-2} s^{-1} (0.1 keV), spectrum $kT = 0.015$–0.035 keV, $N_H = 2 \times 10^{20}$ cm^{-2}. Optics: white dwarf HZ 43, $V = 12.86$, spectrum DA, distance 30–150 pc, $T = 57\,000$ K. (Hearn *et al.*, 1976b; Margon *et al.*, 1976).
13144+595	Forman *et al.* (1978).
13149–646	Forman *et al.* (1978).
13175+067	Forman *et al.* (1978).
13200–105	X-ray: soft burst, in the maximum (0.2–0.4 keV) $\sim 7.5 \times 10^{30}$ erg s^{-1} (?). Optics: flare star dK5e, distance 18.9 pc (?). (Lategan, 1978).

Table II (continued)

13223 − 427	Peculiar galaxy Cen A. X-ray: point source (in center Cen A) and extended by size 2.'8 8.6.75 increase of intensity during 2–4 h. Luminosity (2–6 keV) ∼ 6 × 10^{40} erg s^{-1}. Optics: galaxy of EOp type, V = 7, distance 4.4 mpc. (Cooke *et al.*, 1978, 1978a; Forman *et al.*, 1978; Lawrence *et al.*, 1977; Schnopper and Delvaille, 1977).
13236 − 620	Forman *et al.* (1978), Seward *et al.* (1976a).
13240 − 625	X-ray: transient, maximum was 10.4.67, kT < 3.5 keV, $\tau_{min.}$ < 44 days, τ_e ∼ 20 days. (Cominsky *et al.*, 1978; Rao *et al.*, 1969).
13255 − 020	Cluster of galaxies Abell 1750 (?). Possibly, there are two sources. (Marshall *et al.*, 1979).
13264 + 119	Forman *et al.* (1978), Ricketts *et al.* (1978).
13268 − 311	Cluster of galaxies SC 1329 − 314 (?). X-ray: spectrum kT = 8.2 (+7.3, −3.0) keV, N_H < 1.3 × 10^{22} cm^{-2}, luminosity (2–6 keV) ∼ 1.8 × 10^{45} erg s^{-1}. (Cooke *et al.*, 1978; Lugger, 1978; Markert *et al.*, 1975b; Markert *et al.*, 1976a; McHardy, 1978; Murray and Willmer, 1976; Mushotzky *et al.*, 1978).
13327 − 336	Marshall *et al.* (1979).
13358 + 402	Cluster of galaxies Abell 1763. (Ricketts, 1978).
13442 − 609	Possibly, is weak-condition of source 13240 − 625 (?). (Forman *et al.*, 1978; Seward *et al.*, 1976a; Villa *et al.*, 1976; Watson *et al.*, 1978).
13448 − 325	Cluster of galaxies SC 1345 − 301 (?). X-rays: luminosity (2–6 keV) ∼ 1.2 × 10^{44} erg s^{-1} (?). (Cooke *et al.*, 1978; Forman *et al.*, 1978; Markert *et al.*, 1975b; McHardy, 1978; Murray and Ulmer, 1976; Lugger, 1978).
13464 − 300	Seyfert galaxy IC 4329 A (?). X-rays: spectral index −0.3 (2–10 keV). Luminosity 8 × 10^{43} erg s^{-1} (2–10 keV). (Bradt, 1978; McHardy, 1978).
13468 + 266	Cluster of galaxies Abell 1795 (?). X-rays: luminosity ∼ 10^{44} erg s^{-1} (2–10 keV). (Cooke *et al.*, 1978; Forman *et al.*, 1978, 1978a; McHardy, 1978; Murray and Ulmer, 1976; Rowan-Robinson and Fabian, 1975; Schwartz, 1978; Schwartz *et al.*, 1979).
13481 + 700	Seyfert Galaxy MKN 279 (?). (Cooke *et al.*, 1978).
13505 + 390	Active galaxy MKN 464 (?). (Marshall *et al.*, 1979).
13522 + 187	X-rays: luminosity (0.15–1.5 keV) ∼ 10^{29} erg s^{-1} (?). Optics: binary system η Boo. $P_{orb.}$ = 484, 173 days. K_1 = 8.41 km s^{-1}, $f(M)$ = 0.0276, e = 0.257, $a \sin i$ = 5.5 × 10^{12} cm, spectrum G0 IV, M_1 = 1.2 $M_\odot$, M_2 = 0.45 $M_\odot$. (Topka *et al.*, 1979).
13539 − 645	Markert *et al.* (1975b, 1977), Seward *et al.* (1976a), Villa *et al.* (1976).
14010 − 452	X-rays: in the range 0.1–0.28 keV; intensity ∼0.1 Cyg Loop. Circ. IAU, 1975, No. 2808.
14044 + 145	Cluster of galaxies Abell 1852, 1849 (?). X-rays: luminosity (2–6 keV) ∼ 2 × 10^{45} erg s^{-1}. (Forman *et al.*, 1978; Ricketts, 1978; Schwartz, 1978).
14069 − 619	Probably, is XRS 14104 − 619 (?). Possibly, is in the connection with gamma-source CG 312–1 (?). (Griffiths *et al.*, 1978; Markert *et al.*, 1975b, 1977, 1978; Maraschi *et al.*, 1977; Seward *et al.*, 1976a; Villa *et al.*, 1976).
14104 − 619	Transient. Possibly, is XRS 14069 − 619 (?). X-ray: spectrum kT > 15 keV, life-time ⩾ 30 days. (Cominsky *et al.*, 1978; Wheaton, 1975).

Table II (continued)

14106−029	Emission galaxy NGC 5506 (?).
	(Cooke *et al.*, 1978; Forman *et al.*, 1978; Griffiths *et al.*, 1979; Markert *et al.*, 1978).
14156+255	Seyfert galaxy NGC 5548 (?).
	(Cooke *et al.*, 1978; Forman *et al.*, 1978).
14162−589	Cluster of galaxies Abell 1415 (?).
	(Watson *et al.*, 1978).
14163−622	Possibly association with gamma-source CG 312−1 (?).
	(Forman *et al.*, 1978; Griffiths *et al.*, 1978; Maraschi *et al.*, 1974).
14174−624	X-ray: spectral index -0.11 ± 0.044.
	(Apparao *et al.*, 1979).
14185−614	Forman *et al.* (1978), Markert *et al.* (1975b, 1977, 1978), Seward *et al.* (1976a), Villa *et al.* (1976).
14186+485	Cluster of galaxies Abell 1904 (?).
	X-ray: luminosity (2–10 keV) $\sim 5 \times 10^{42}$ erg s^{-1}.
	(Cooke *et al.*, 1978; McHardy, 1978).
14260−624	Proxima Cen.
	X-ray: luminosity in maximum of burst (0.05–0.29 keV) 10^{30} erg s^{-1}, in normal condition is less than 2.2×10^{28} erg s^{-1}.
	Optics: $V = 11.05$, spectrum M5, mass 0.1 $M_\odot$, distance 1.3 pc.
	(Haisch *et al.*, 1978).
14362−606	α Cen.
	X-ray: in the range ~ 0.25 keV, intensity 1.6×10^{-11} erg cm^{-2} s^{-1}, luminosity $\sim 3 \times 10^{27}$ erg s^{-1}, spectrum $kT \sim 0.035$ keV.
	Optics: visual-binary system, $V_1 = -0.01$, $V_2 = 1.33$, spectrum G2 V + K5 V, distance 1.34 pc.
	(Nugert and Garmire, 1978).
14365−566	Forman *et al.* (1978).
14369−620	Possibly the source XRS 14392−614 (?).
	(Forman *et al.*, 1977; Hill *et al.*, 1972).
14384−185	Forman *et al.* (1978).
14390−617	Possibly the source XRS 14369−620 (?).
	(Seward *et al.*, 1976a; Villa *et al.*, 1976; Watson *et al.*, 1978).
14391−622	Supernova remnant MSH 14–63.
	X-rays: $T_1 (\sim 0.2 \text{keV}) \sim (2.5–7) \times 10^6$ K; $T_2 (> 5 \text{keV}) > 6 \times 10^7$ K; $N_H = (0.5–1) \times 10^{22}$ cm^{-2}; luminosity in the range 2–10 keV $\sim 1.4 \times 10^{35}$ erg s^{-1}.
	Radio: burst 185 yr (?). Distance 2.5 kpc, $Z \simeq 100$ pc.
	(Culhane, 1977; Markert *et al.*, 1978; Naranan *et al.*, 1976; van den Bergh, 1978; Winkler, 1978; Winkler, 1978).
14446+430	Bahcall *et al.* (1976b), Cooke *et al.* (1978), Forman *et al.* (1978), Markert *et al.* (1978), Murray and Ulmer (1976).
14484−556	Transient source (?) with large flare duration.
	(Amnuel *et al.*, 1974; Chodil *et al.*, 1967; Cruddace, 1971; Forman *et al.*, 1978; Friedman *et al.*, 1967; Hill *et al.*, 1972; Lewin *et al.*, 1968a; Seward, 1978).
14490+193	Star ζ Boo.
	X-ray: luminosity (0.2–2.8 keV) $\sim 2.4 \times 10^{29}$ erg s^{-1}.
	Optics: binary system, emission stars, K5 IV + G8 IV.
	$P_{orb.} = 149.95$ yr, $V = 4.64$.
	(Woltjer, 1978a).
14506−805	Forman *et al.* (1978).
14528−602	Forman *et al.* (1976a), Seward *et al.* (1976a).
14450+191	Cluster of galaxies Abell 1991 (?).
	(Forman *et al.*, 1978; Ricketts, 1978).
14553−313	Transient source:
	X-ray: spectrum $kT \simeq 3$ keV. Two bursts (1969 and 1979 yrs).
	burster, burst time $\sim$ s.
	Optics: star, increase of brightness from 19 to 13 magnitude 10.5.79: $V = 12.9$, $B - V = 0.0$, $U - B = -0.8$; 14.6.79: $B \sim 18$.

Table II (continued)

	Radio: burst 28.5.79, $S_{max} \sim 0.008$ Jy.
	(Conner *et al.*, 1969; Evans *et al.*, Z970; Kudriavtsev *et al.*, 1976; Seward and Liller, 1977; van Genderen, 1979).
14554−273	Forman *et al.* (1978).
14566+225	Forman *et al.* (1978), Ricketts (1978).
14580−415	Supernova remnant 1006 yr.
	X-ray: spectrum $kT = 4$ keV, $N_H < 4 \times 10^{21}$ cm^{-2}.
	Luminosity (2–6 keV) $\sim 1.4 \times 10^{34}$ erg s^{-1}.
	Optics: filamentar nebula.
	Radio: PKS 1459−41, diam 25–30 pc, distance 1.3 kpc, spectr. index $\alpha = -0.57$.
	(van den Bergh, 1976; Cooke *et al.*, 1978; Culhane, 1977; Forman *et al.*, 1978; Gorenstein and Tucker, 1975; Markert *et al.*, 1978; Zarnetzky and Bibbo, 1979; Sinkler *et al.*, 1978, 1979; Winkler, 1978; Winkler and Laird, 1976; Woltjer, 1972).
15059+573	Forman *et al.* (1978).
15087+062	Cluster of galaxies Abell 2029 (?).
	Possibly, is source 15142+068 (?).
	X-ray: spectrum $kT = 6.2$ (+2.6; −1.6) keV, $N_H < 7.7 \times 10^{21}$ cm^{-2}, luminosity (2–6 keV) $\sim 1.8 \times 10^{45}$ erg s^{-1}.
	(Cooke *et al.*, 1978; McHardy, 1978; Mushotzky *et al.*, 1978; Schwartz, 1978; Seward *et al.*, 1979a).
15100−390	Supernova remnant Lup Loop.
	X-ray: spectrum $kT = 0.22$ keV, $N_H = 7 \times 10^{20}$ cm^{-2}; in the range 0.2–1 keV, intensity $\sim 3 \times 10^{-10}$ erg cm^{-2} s^{-1}, luminosity (0.15–0.8 keV) $\sim 5 \times 10^{34}$ erg s^{-1}.
	Radio: on 408 MHz, flux 445 f.u., distance 0.5 kpc, diam 48.6 pc.
	(Culhane, 1977; Hill *et al.*, 1972; Palmieri *et al.*, 1979; Winkler, 1978a; Winkler *et al.*, 1979).
15101−590	Supernova remnant MSH 15−52 AB (?).
	X-ray: luminosity (2–6 keV) $\sim 1.5 \times 10^{36}$ erg s^{-1}.
	Radio: diam 24 pc. distance 3.2 kpc.
	(Culhane, 1977; Forman *et al.*, 1978; Mc-Connell and Cowley, 1972; Markert *et al.*, 1978; Seward *et al.*, 1976a).
15134+070	Probably, is source 15134+070 (?).
	Cluster of galaxies Abell 2052 (?).
	(Marshall *et al.*, 1979).
15142+068	Probably, is source 15142+068 (?).
	(Markert *et al.*, 1976a, 1978).
15150+231	Forman *et al.* (1978).
15168−569	X-ray: strong variability and bursts, spectrum $kT > 15$ keV.
	X-ray: pulsations were not observed, $P_{orb.} = 16.585 \pm 0.01$ days (?).
	Optics: emission object $B \sim 22.5$, IR-region: $K = 7.66–9.16$, $J = 9.88$, $P_{orb.} = 16.595$ days.
	(Bradt *et al.*, 1977b; Berezhnoy *et al.*, 1977; Clark *et al.*, 1975; Cruddace, 1971; Davison and Tuohy, 1975; Forman *et al.*, 1976c, 1978; Glass, 1976; Glass, 1978, 1979; Haynes *et al.*, 1976; Haynes *et al.*, 1978; Hill *et al.*, 1972; Joss *et al.*, 1976; Johns *et al.*, 1974; Johns, 1976; Kaluzienski and Holt, 1978a; Kaluzienski *et al.*, 1976c; Margon *et al.*, 1971; Markert *et al.*, 1978; Seward, 1970; Seward *et al.*, 1972; Saden *et al.*, 1979; Samimi *et al.*, 1979; Toor, 1977; Villa *et al.*, 1976; Wilson and Carpenter, 1976).
15170−500	Duration of burst is less than 128 s.
	Probably, is XRS 15168−569 (?).
	(Evans *et al.*, 1976).
15190−314	Duration of burst is < 128 s.
	Possibly, burst was precursor of flare XRS 14553−313 (?).
	(Evans *et al.*, 1976).
15190+082	Possibly, NGC 5920 (??).
	(Cooke *et al.*, 1978).
15212+285	Cluster of galaxies Abell 2065 (?).
	X-ray: luminosity (2–6 keV) $\sim 4 \times 10^{44}$ erg s^{-1}.
	(Cooke *et al.*, 1978; Forman *et al.*, 1978; McHardy, 1978).

Table II (continued)

15241 − 617	Transient source.

15241 − 617 Transient source.
X-ray: time of increase ~19 days, decrease ~30 days, spectrum $kT = 1.4$ keV, $N_H = 7.1 \times 10^{21}$ cm^{-2}.
Optics: star $B = 17.5$ (15.12.74), decrease of brightness to $B = 22$ (3.4.76), $M_{bol.} \geqslant 6$ (red dwarf?).
(Bradt *et al.*, 1977b; Carpenter *et al.*, 1977; Kaluzienski *et al.*, 1975b).

15241 − 617 Bradt *et al.* (1978), Maraschi *et al.* (1976), Markert *et al.* (1977), Murdin *et al.* (1977).

15307 − 443 Forman *et al.* (1978).

15358 − 292 One flare with the duration less than 25 min.
(Forman *et al.*, 1978).

15382 − 521 X-ray: pulsar $P = 528.929 \pm 0.040$ s, $(-\dot{P}/P) \leqslant 9.7 \times 10^{-4}$ yr^{-1}, $N_H \simeq (1-2) \times 10^{22}$ cm^{-2},
luminosity $\leqslant 1.8 \times 10^{36}$ erg s^{-1}, spectrum hard, $K_x = 323$ km s^{-1}.
Optics: star $V = 14.35$, $B - V = 1.91$, $U - B = -0.58$, spectrum B0 I, binary system $P_{orb.}$
$= 3.7918$ days, $K_0 = 33$ km s^{-1}, $f(M) = 13\ M_\odot$, distance less than 6.5 kpc, masses of components
1.5–2 $M_\odot$ (neutron star) and ~19 $M_\odot$ (normal star).
(Apparao *et al.*, 1978; Bradt *et al.*, 1978; Bradt, 1978; Cruddace *et al.*, 1972; Cowley *et al.*, 1978;
Crampton, *et al.*, 1978; Davison *et al.*, 1977; Forman *et al.*, 1978; Glass *et al.*, 1979; Ilovaisky *et al.*,
1979; Kaluzienski and Holt, 1978; Markert *et al.*, 1978; Parkes *et al.*, 1978a; Schwartz *et al.*, 1978;
Seward *et al.*, 1976a).

15400 − 320 X-ray: probably, is old supernova remnant, with radioremnants is not identified.
In the range 0.15–1.8 keV flux 5.3×10^{-10} erg cm^{-2} s^{-1}, spectrum $kT = 0.2$ keV, $N_H = 10^{21}$
cm^{-2}.
(Agrawal and Riegler, 1978).

15412 − 534 Probably, is source 15382 − 521 (?).
(Becker *et al.*, 1977a; Davison, 1977b; Watson and Pye, 1977).

15430 − 624 X-ray: spectrum $kT = 3$–7 keV, $E_a \leqslant 2.6$ keV.
Optics: star $B \sim 20$.
(Apparao *et al.*, 1978; Bradt *et al.*, 1978; Forman *et al.*, 1978; Johns, 1976; Markert *et al.*, 1978;
Seward *et al.*, 1976a; Willmer *et al.*, 1974).

15438 − 475 X-ray: transient. Value of kT changed from 2.1 keV (30.10.71) to 0.9 keV (28.4.72). Irregular
intervals of high and low luminosity conditions were observed.
$\tau_{min.} > 110$ days, $\tau_e \sim 10$ days, $kT < 3$ keV.
Optics: $K = 8.56$; $\Delta K \approx 0.2$.
(Cominsky *et al.*, 1978; Forman *et al.*, 1978; Forman and Liller, 1973; Glass *et al.*, 1979; Johns,
1976; Kudriavtsev *et al.*, 1976; Li *et al.*, 1976b; Matilsky *et al.*, 1972; Villa *et al.*, 1976).

15440 − 757 Markert *et al.* (1978).

15452 − 536 Supernova remnant MSH 15–56.
X-ray: $kT \sim 0.8$ keV, $N_H \sim 10^{22}$ cm^{-2}, in the range 0.2–2 keV flux 5.4×10^{-11} erg cm^{-2} s^{-1}.
Radio: on frequence 1 GHz flux 26 f.u. spectral index $\alpha = -0.78$.
(Agrawal and Riegler, 1978a).

15539 − 542 Transient source. IR-region: star $K = 12.0$, gamma-region: source CG 327–0 (?).
(Apparao *et al.*, 1978; Bradt *et al.*, 1978; Glass, 1979; Griffiths *et al.*, 1978; Walker, 1976).

15561 − 756 Cooke *et al.* (1978).

15564 − 527 Seward *et al.* (1976a).

15565 + 272 Cluster of galaxies Abell 2142 (?).
X-ray: spectrum $kT \geqslant 33$ keV, $N_H < 1.4 \times 10^{22}$ cm^{-2}, luminosity (2–10 keV) ~ $(2-6) \times 10^{45}$
erg s^{-1}.
(Cooke *et al.*, 1978; Forman *et al.*, 1978a; Markert *et al.*, 1978; McHardy, 1978; Mushotzky *et al.*,
1978; Nitzen and Scott, 1979; Schwartz, 1978).

15569 − 606 Apparao *et al.* (1978), Forman *et al.* (1978), Class (1979), Harries *et al.* (1971), Markert *et al.* (1978),
Seward *et al.* (1976a), Willmore *et al.* (1974).

16011 + 159 Cluster of galaxies Abell 2147, 2152 (?).
X-rays: spectrum $kT = 7.2$ (+1.4 − 1.1) keV, $N_H = 7.2 \times 10^{21}$ cm^{-2}, luminosity (2–10 keV)
~ 3.1×10^{44} erg s^{-1}.
(Cooke *et al.*, 1978; Forman *et al.*, 1978; McHardy, 1978; Mushotzky *et al.*, 1978; Schwartz, 1978).

Table II (continued)

16040 − 590 X-rays: many bursts.
(Belian *et al.*, 1976c).

16088 − 522 X-rays: steady source, spectrum $kT = 7$–15 keV, $N_H = 2.5 \times 10^{22}$ cm^{-2}, burst of duration
30–100 s, transient phenomenon $\sim$ months.
Optics: star $V = 18.2$ (August, 1977), out of burst $V \geqslant 23$–24.
(Apparao *et al.*, 1978; Belian *et al.*, 1976c; Clark *et al.*, 1977d; Cominsky *et al.*, 1978; Grindlay and
Gursky, 1976a; Grindlay and Liller, 1978; Holt, 1978b; Hill *et al.*, 1972; Johns, 1976; Kaluzienski
and Holt, 1978b, c, and d; Lewin, 1977; Liller, 1978; Li, 1976; Levin *et al.*, 1976a; Markert *et al.*,
1978; Markert *et al.*, 1977; Oda, 1978; Ricker *et al.*, 1976a; Seward *et al.*, 1976a; Swank *et al.*, 1976b;
Tananbaum *et al.*, 1976).

16128 + 339 X-rays: luminosity (0.2–2.8 keV) $\sim 4.6 \times 10^{30}$ erg s^{-1}, spectrum $kT \sim 0.8$ keV.
Optics: star σ CrB, type RS CVn-binary system, spectrum F6V + GI, $P_{orb.} = 1.14$ days,
$V = 5.76$.
(Bopp and Talcott, 1978).

16139 − 509 Supernova remnant RCW 103.
X-ray: spectrum $kT = 0.2$ keV, $N_H = 9 \times 10^{21}$ cm^{-2}, luminosity (2–10 keV) $\approx 10^{25}$ erg s^{-1}, age
20 000 yr, $V = 400$ km s^{-1}.
Radio: 1 GHz flux 25 f.u. spectr. index -0.55, distance 3.3–8 kpc.
(Tuohy *et al.*, 1979, 1978).

16140 − 395 Cruddace (1971), Fisher *et al.* (1968), Friedman *et al.* (1967), Fujimoto *et al.* (1969), Meyer *et al.*
(1970), Seward *et al.* (1970).

16146 − 277 Forman *et al.* (1978).

16170 − 155 X-ray: two conditions, active (bursts) and passive. In the active condition spectrum $kT \sim 5.5$–20
keV (between bursts $kT = 5.5$ keV). In the passive condition spectrum $kT \sim 7.5$ keV, variability
20–50%, $N_H \sim 6 \times 10^{20}$–1.3 $\times 10^{23}$ cm^{-2}, luminosity $\sim 2 \times 10^{37}$ erg s^{-1} (2–6 keV, distance 0.7
kpc).
Optics: V 818 Sco, $B = 12.2$–12.4. Optical bursts correlate with the X-ray ones. Strong emission
lines He II 4686, N III–C III 4640–4650. $P_{orb.} = 0.787$ days, $V_c = -138.5$ km s^{-1}. Masses of
components nearly 2 $M_\odot$, distance 0.3–1 kpc.
(Basko *et al.*, 1976; Berezhnoy *et al.*, 1977; Chi-Chao, 1979; Cowley and Crampton, 1975;
Crampton *et al.*, 1976b; Culhane *et al.*, 1976; Goven *et al.*, 1977; Glass, 1979; Holt *et al.*, 1976a;
Johns, 1976; Long and Kestenbaum, 1978; Markert *et al.*, 1978; Moskalenko *et al.*, 1976; Petro,
1978; Parsignault and Grandlay, 1978; Wite *et al.*, 1975a).

16186 + 150 Supernova remnant North Polar Spure.
X-ray: in the region 0.18–0.28 keV intensity 1.8×10^{-8} erg cm^{-2} s^{-1}, distance 140 pc, lumi-
nosity (0.1–1 keV) 3×10^{36} erg s^{-1}. Age (1.4–3.8) $\times 10^5$ yr.
(Cruddace *et al.*, 1976; Hayakawa *et al.*, 1977).

16190 − 468 X-ray: burst with duration less than 128 s.
(Evans *et al.*, 1976).

16210 − 528 Seward *et al.* (1976a).

16212 − 234 Cluster of galaxies (?).
(Forman *et al.*, 1978).

16220 − 243 X-ray: spectral index $\alpha = -0.4$, $N_H > 10^{22}$ cm^{-2}, extended source $\sim 1°5$, size of cloud nearly
5 pc.
Density of cloud 700–10^4 cm^{-3}, distance 160–200 pc.
(Apparao *et al.*, 1979a).

16240 + 659 Active galaxy MKN 576 (?).
(Marshall *et al.*, 1978).

16243 − 490 Transient source (?).
X-ray: maximum in 1965 yr. Vela-5 fixed the bursts from this region (May, 1969–August, 1970);
spectrum $kT = 3$–7 keV.
Optics: in the range – open cluster NGC 6134.
(Apparao *et al.*, 1978; Belian *et al.*, 1976c; Bradt *et al.*, 1978; Fisher *et al.*, 1968; Forman *et al.*, 1978;
Friedman *et al.*, 1967; Hill *et al.*, 1972; Johns, 1976; Markert *et al.*, 1978; Parsignault and
Grandlay, 1978; Sanduleak and Dolan, 1974; Villa *et al.*, 1976; Willmore *et al.*, 1974).

Table II (continued)

16256 − 333	Forman *et al.* (1978).
16270 − 092	Forman *et al.* (1978).
16272 − 673	X-ray: pulsar $P = 7.6806$ s, spectrum hard, index 0.3 ± 0.4 orbital period $P_{orb.} = 0.1$ days. Optics: star $V = 18.5$–19.5, pulsar $P = 7.681$ s. (Bradt *et al.*, 1977b; Cooke *et al.*, 1978; Forman *et al.*, 1978; Grindlay, 1978b; Joss *et al.*, 1978; Ilovaisky *et al.*, 1978a; Ilovaisky *et al.*, 1978b; Markert, 1976a; Markert *et al.*, 1978; McClintock *et al.*, 1977b; Seward *et al.*, 1976a).
16278 + 396	Cluster of galaxies Abell 2199 (?). X-ray: spectrum $kT = 3.2 (+1.7; -1.0)$ keV, luminosity (2–6 keV) $\sim 2.7 \times 10^{44}$ erg s^{-1}. (Cooke *et al.*, 1978; Forman *et al.*, 1978; Henry and Tucker, 1979; McHardy, 1978; Mushotzky *et al.*, 1976; Schwartz, 1978; Schwartz *et al.*, 1979a).
16284 + 286	Cluster of galaxies Abell 2200 (?). X-ray: luminosity (2–6 keV) $\sim 3.9 \times 10^{45}$ erg s^{-1}. (Cooke *et al.*, 1978; Forman *et al.*, 1978; Schwartz, 1978).
16301 − 472	Transient source. X-ray: bursts with intervals ~ 600 days, spectrum $kT = 2$ keV, duration of burst ~ 50 days. Optics: star $V = 15.5$. (Cruddace *et al.*, 1972; Forman *et al.*, 1976a; Grindlay, 1977).
16301 − 472	Cominsky *et al.* (1978), Holt *et al.* (1978), Johns (1976), Kaluzienski and Holt (1977), Markert *et al.* (1978), Sims and Watson (1978), Share *et al.* (1978a), Villa *et al.* (1976).
16315 − 643	Cooke *et al.* (1978), Forman *et al.* (1978), Markert *et al.* (1978), Seward *et al.* (1976a), Webster (1973).
16364 + 052	Cluster of galaxies Abell 2204 (?). X-ray: luminosity (2–6 keV) $\sim 1.3 \times 10^{45}$ erg s^{-1}. (Cooke *et al.*, 1978; Forman *et al.*, 1978; Markert *et al.*, 1976a; McHardy, 1978).
16369 − 536	X-ray: X-ray burst with duration less than ~ 10 s and intervals nearly 3.3 h (OSO-8) and ~ 11 h (SAS-3). Optics: blue star with $V = 17.52$, $B - V = 0.69$, $U - B = 0.70$, $B = 18.21$; Emission line $\lambda 4630$–40 (N III) and $\lambda 4686$ (He II). $F_{opt.} = 7.1 \times 10^{-13}$ erg cm^{-2} s^{-1}. Doppler velocity 1000 km s^{-1}. Spectr. variability from $\sim$ hours to $\sim$ days. (Canizares *et al.*, 1979; Forman *et al.*, 1978; Friedman *et al.*, 1967; Hoffman, 1977; Johns, 1976; Markert *et al.*, 1978; McClintock *et al.*, 1977b; Murdin *et al.*, 1974; Parsignault and Grindlay, 1978; Seward *et al.*, 1972, 1976a; Swank *et al.*, 1976b; Thomas *et al.*, 1975; van Paradijs *et al.*, 1979; Willmore *et al.*, 1974).
16393 + 402	Markert *et al.* (1978).
16400 + 400	Soft source. Spectrum softer, than spectrum Cyg Loop. (Clark, 1975a).
16408 − 463	Hill *et al.* (1972).
16410 − 326	Marshall *et al.* (1979).
16420 − 343	Many bursts with duration ~ 1 s. (Belian *et al.*, 1976c).
16421 − 455	X-ray: spectrum $kT = 2$–5 keV. (Apparao *et al.*, 1978; Cruddace, 1971; Forman *et al.*, 1978; Fisher *et al.*, 1968; Friedman *et al.*, 1967; Fujimoto *et al.*, 1969; Glass, 1979; Meyer *et al.*, 1970; Parsignault and Grindlay, 1978; Seward, 1970; Seward *et al.*, 1976a; Thomas *et al.*, 1975).
16446 + 699	Forman *et al.* (1978).
16450 − 575	Many bursts. (Belian *et al.*, 1976c).
16450 − 507	Burst with duration less than 128 s. (Evans *et al.*, 1976).
16450 − 284	Marshall *et al.* (1979).
16460 + 284	Burst with duration less than 128 s. (Evans *et al.*, 1976).
16489 − 185	Marshall *et al.* (1979).
16494 − 595	Marshall *et al.* (1979).

Table II (continued)

16516+399 Galaxy MKN 501 (?) type.
X-ray: luminosity (2–6 keV) $\sim 1.6 \times 10^{44}$ erg s^{-1}.
Optics: $B = 14.5$, variability ~ 2.5.
X-ray: spectral index -0.2 (0.1–1.0 GHz).
(Forman *et al.*, 1978; Schwartz *et al.*, 1978c).

16518−065 Forman *et al.* (1978).

16528+635 Forman *et al.* (1978).

16531−407 X-ray: spectrum $kT = 11 \pm 4$ keV, $N_H \simeq 4 \times 10^{23}$ cm^{-2}, pulsar $P = 38.22$ s (5.5–13.5 keV).
Optics: V 861 Sco (identification, probably, is not right).
$V = 6.2$, spectrum B0. 5Ia, distance 1.4 kpc; $P_{orb.} = 7.848$ days, eclipses.
(Glass, 1977; Polidan *et al.*, 1978; Polidan *et al.*, 1978a; Pacheko, 1978; White *et al.*, 1978b).

16559−424 Transient source (?).
X-ray: spectrum $kT = 3.2 \pm 2.7$ keV.
(Fisher *et al.*, 1968; Hill *et al.*, 1972; Seward, 1970).

16560+354 X-ray: pulsar $P = 1.2378$ s, binary system.
$P_{orb.} = 1.700\,165$ days. Period 'on' and 'of' is equal 35.7 days.
Spectrum has soft ($T \simeq 4.5 \times 10^5$ K) and hard ($kT > 15$ keV) components.
Optics: star HZ Her, $V = 12$–14, spectrum A7–B0, masses of components ~ 2(norm.) and 1.3 (X-rays) $M_\odot$.
(Avni and Bahcall, 1976; Berezhnoy *et al.*, 1977; Bunner, 1978; Becker *et al.*, 1977b; Bernacca, 1976; Brecher and Wasserman, 1974; Burke, 1976; Catura and Acton, 1976b; Cooke *et al.*, 1978; Crampton and Cowley, 1976; Fechner and Joss, 1977; Forman *et al.*, 1978; Fritz *et al.*, 1976; Holt *et al.*, 1979; Henry and Schreier, 1977; Heise and Brinkman, 1976; Joss *et al.*, 1974; John, 1977; Johns, 1976; Joss *et al.*, 1977; Lumb *et al.*, 1976; Liller, 1976a; Markert *et al.*, 1978; Middleditch and Nelson, 1975; Oke, 1976; Pravdo *et al.*, 1978; Pravdo *et al.*, 1978a; Pravdo *et al.*, 1977; Shulman *et al.*, 1976; Trümper *et al.*, 1978).

16589−298 X-rays: 24 bursts with duration 30 s and intervals 2.2–2.6 h, kT 4 keV, $N_H = 2 \times 10^{22}$ cm^{-2}.
Optics: star $V = 18.3$, $B - V = 0.37$; $U - B = -0.38$, emission N III, He II.
(Canizares and McClintock, 1979; Doxey *et al.*, 1979; Grindlay, 1978c; Lewin *et al.*, 1976b; van Paradijs, 1979).

16589−487 X-rays: spectrum kT 3 keV, luminosity more than 2×10^{37} erg s^{-1}.
Optics: $V = 16.6$, $B - V = 0.84$, $U - B = -0.24$, distance more than 4 kpc.
(Forman *et al.*, 1978; Grindlay, 1978d; Doxsey *et al.*, 1979; Johns, 1976; Li *et al.*, 1978a; Markert *et al.*, 1979; Seward *et al.*, 1976a).

16592+337 Cluster of galaxies Abell 2244/2245 (?).
X-rays: luminosity 1.6×10^{45} km s^{-1}.
(Cooke *et al.*, 1978; Forman *et al.*, 1978; McHardy, 1978).

16594−765 Forman *et al.* (1978), Giacconi *et al.* (1974).

17005−377 X-rays: binary system, $P_{orb.} = 3.4118$, eclipses, spectrum $kT > 15$ keV, variability with period 97 min, $K_x = 400$ km s^{-1}.
Optics: star HD 153919, $V = 6.75$, $V = 0.04$, spectrum O6f, $K_{opt.} = 19$ km s^{-1}, masses of components 25 $M_\odot$ (norm.) and 1.3 $M_\odot$ (X-ray source), oscillations with period 95 $\pm$ 3 min.
(Branduardi *et al.*, 1978; Dupree and Lester, 1976; Forman *et al.*, 1978; Hutchings, 1978; Hammerschlag-Hensberge and Wu, 1977; Jernigan *et al.*, 1978; Jernigan *et al.*, 1978; Jernigan *et al.*, 1978a; Joss, 1976; Kondo *et al.*, 1976; Markert *et al.*, 1978; Matilsky and Jessen, 1978; Matilsky *et al.*, 1978; Mason *et al.*, 1976b; Perry and Peterson, 1974; Surdej, 1978; Walker, 1976).

17023−363 X-ray: spectrum $kT = 5.8$ keV, $E_a \leqslant 2.6$ keV, $N_H \approx 10^{22}$ cm^{-2}.
Optics: star $B > 21.5$, $V \sim 19$ (?). Probably, late M-giant.
(Bowyer *et al.*, 1965; Cruddace *et al.*, 1972; Fisher *et al.*, 1978; Fujimoto *et al.*, 1969; Glass and Feast, 1978; Glass, 1979; Gorenstein, 1969; Hill *et al.*, 1972; Jones, 1976; Jernigan *et al.*, 1978; Jernigan *et al.*, 1978a; Meyer *et al.*, 1970; Markert *et al.*, 1978; McClintock and Canizares, 1978; Parsignault and Grindlay, 1978; Seward, 1970; Seward *et al.*, 1972; White *et al.*, 1975; Zuiderwijk, 1978; Zuiderwijk, 1975a).

17023−429 X-ray: spectrum $kT = 7$–15 keV, $E_a \leqslant 2.6$ keV.
(Forman *et al.*, 1978; Hill *et al.*, 1972; Jernigan *et al.*, 1978; Thomas *et al.*, 1975; Villa *et al.*, 1976).

Table II (continued)

$17036+261$	Forman *et al.* (1978).
$17040+241$	Cooke *et al.* (1978), Forman *et al.* (1978).
$17043-304$	Forman *et al.* (1978).
$17045-320$	Markert *et al.* (1978).
$17051-431$	X-ray: 5 bursts of duration 20–30 s. Intervals from 0.006 to 190 days. (Swank *et al.*, 1976a).
$17051-250$	Transient source. X-ray: spectral index ~ -0.4, $N_H \approx 3 \times 10^{21}$ cm^{-2}. Optics: star $B = 16.5$ (during burst) $B = 21$ (before burst), $A_v \sim 1.4$, emission He II 4686. (Griffiths *et al.*, 1977; Griffiths *et al.*, 1978; Griffiths *et al.*, 1978b).
$17052+609$	Cooke *et al.* (1978).
$17054-440$	X-ray: spectrum $kT = 3$–7 keV, $E_a \leqslant 2.6$ keV. (Forman *et al.*, 1976c, 1978; Jernigan *et al.*, 1978; Jones, 1976; Markert *et al.*, 1978; Thomas *et al.*, 1975).
$17056-322$	Forman *et al.* (1978).
$17064+321$	Markert *et al.* (1978).
$17065-273$	Globular cluster NGC 6293 (?). (Pye *et al.*, 1976).
$17068-434$	OSO-8 and SAS-3 observed 5 bursts. $\tau_e \sim 20$–30 s interval between bursts from 0.006 to 190 days. (Marshall *et al.*, 1978; Swank *et al.*, 1976).
$17074+786$	Cluster of galaxies Abell 2256 (?). X-ray: spectrum $kT = 7.4$ keV, $N_H < 1.3 \times 10^{22}$ cm^{-2}, luminosity (2–6 keV) $\sim 9.5 \times 10^{44}$ erg s^{-1}. (Cooke *et al.*, 1978; Forman *et al.*, 1978; Henry and Tucker, 1979; Markert *et al.*, 1978; McHardy, 1978; Mushotzky *et al.*, 1978; Schwartz, 1978).
$17083-407$	X-ray: spectrum $kT = 3$–7 keV, $E_a \leqslant 2.6$ keV. (Forman *et al.*, 1978; Jones, 1976; Markert *et al.*, 1975b, 1977; Markert *et al.*, 1978).
$17089-232$	X-ray: spectrum $kT = 3$–7 keV, $E_a \leqslant 2.6$ keV. Optics: star $V = 18.2$, Balmer absorbtion lines, blue continuum. (Davidsen *et al.*, 1976; Forman *et al.*, 1978; Friedman *et al.*, 1967; Jones, 1976; Markert *et al.*, 1978).
$17100-510$	Many bursts of duration 1 s. (Belian *et al.*, 1976c).
$17104-303$	During burst kT increases from 1.8 keV to 9.5 keV and decreases to 3.2 keV. Weak source has $kT = 10 \pm 2$ keV. (Swank *et al.*, 1977).
$17106-474$	Hard source. (Bratoliubova-Tsulukidze *et al.*, 1976b).
$17108-340$	Possibly, is XRS $17196-347$ (?). (Carpenter *et al.*, 1977).
$17151-393$	Markert *et al.* (1978), Sanduleak and Dolan (1974).
$17154+028$	Forman *et al.* (1978).
$17160-487$	X-ray: spectrum $kT = 1.7$ keV, $E_a = 1.5$ keV, $N_H = 10^{22}$ cm^{-2}. (Bunner and Palmieri, 1969).
$17166-016$	Forman *et al.* (1978).
$17161-318$	X-ray: bursts of duration 5–10 min, spectrum $kT = 17$ keV, $N_H = 6 \times 10^{21}$ cm^{-2}. (Jernigan *et al.*, 1978; Markert *et al.*, 1976b, 1977; Markert *et al.*, 1978).
$17194+661$	Cluster of galaxies Abell 2255 (?). (Cooke and Maccagni, 1976; Cooke *et al.*, 1978; Ricketts, 1978).
$17196-347$	Possibly is XRS $17108-340$ (?). X-ray: transient (?). (Fisher *et al.*, 1968; Greenstein, 1969; Hill *et al.*, 1972; Seward, 1970).
$17202+346$	Cluster of galaxies Abell 2261, 2266 (?). (Forman *et al.*, 1978; Ricketts, 1978).
$17228-305$	Possibly, is source $17250-302$ (?). X-rays: burst of duration ~ 100 s, $N_H \approx 9 \times 10^{21}$ cm^{-2}, $kT = 10 \pm 2$ keV, $L_x \approx$

Table II (continued)

	2×10^{36} erg s^{-1} (in steady state) $L_x \approx 3\,(+4, -2) \times 10^{38}$ erg s^{-1} (burst).

Optics: Globular cluster Terzian 2 (?) $m_R = 16$, $D = 7 \pm 3$ kpc, radius of core ≈ 0.2 pc, centr. density 10^4–$10^5\, M_\odot$ pc^{-3}.
(Forman *et al.*, 1978; Grindlay, 1978).

17228 + 119   Forman *et al.* (1978).

17250 − 302   Possibly, is 17228 − 305 (?).
X-rays: transient (?) $kT = 4$ keV, $N_H = 3.5 \times 10^{21}$ cm^{-2} (14.6.69).
(Cruddace, 1971; Cruddace *et al.*, 1972; Hill *et al.*, 1972).

17270 − 335   Sequence of bursts with intervals 17 s and tails 1–3 min. Duration of burst ~1 s.
(Markert *et al.*, 1978; Mason *et al.*, 1976a).

17271 + 503   X-rays: the presence of source is doubtful.
Optics: object IZw 1727 + 502, BL Lac. type, $M_B = 16.9$, optical variability 5.9.
(Schwartz *et al.*, 1978b).

17276 − 214   SNR Kepler (1604) = V 843 Oph.
X-rays: flux (0.5–1.5 keV) constant during observations.
(Bunner, 1978).

17286 − 337   X-rays: observ. steady and burst states, duration of bursts ~10 s, intervals ~3–8 hr. Steady source has $kT = 7$–15 keV, during bursts $kT = 2$ keV. If emission of black body then radius of emission region 6 km, distance to the source ~5 kpc, $N_H = (2.3 \pm 0.4) \times 10^{22}$ cm^{-2}.
(Forman *et al.*, 1978; Hoffman *et al.*, 1979; Hoffman *et al.*, 1976a, 1977; Hill *et al.*, 1972; Jones, 1976; Lewin, 1976a; Lewin *et al.*, 1976a, b; Lewin, 1977; Parsignault and Grindlay, 1978; Penni *et al.*, 1975; Thomas *et al.*, 1975; van Paradijs *et al.*, 1979; Willmore *et al.*, 1974).

17288 − 162   X-ray: $kT = 3$–7 keV, $E_a \leqslant 2.6$ keV. Positive correlation between intensity and spectral hardness is observed.
(Bradt *et al.*, 1971; Davidsen *et al.*, 1976; Forman *et al.*, 1978; Jones, 1976; Kasturirangan *et al.*, 1976; Lewin *et al.*, 1968b; Markert *et al.*, 1978; Meyer *et al.*, 1970; Parsignault and Grindlay, 1978; Rappaport *et al.*, 1969; Reina, 1974; White *et al.*, 1975; Willmore *et al.*, 1974).

17289 − 247   X-ray: X-ray pulsar $P = 122.607 \pm 0.006$ s, $kT = 18 \pm 12$ keV.
Optics: star with $V = 18.7$, strong emission Hα, Hβ, He I 5876.
(Becker *et al.*, 1976; Davidsen *et al.*, 1976; Doty, 1976; Glass and Feast, 1973; Glass, 1979; Hawkins and Sanford, 1976; Jones, 1976; Markert *et al.*, 1978; Parsignault and Grindlay, 1978; Ricker *et al.*, 1976; Thomas *et al.*, 1975; White *et al.*, 1975).

17300 − 370   Flux in the range (0.37–1.9) keV $\approx 2 \times 10^{-9}$ erg cm^{-2} s^{-1}. $L_x = 4.8 \times 10^{31}$ erg s^{-1}.
Optics: $\alpha = 17^h 9^m$, $\delta = -43°27'1$, type of star dK7e. Distance 13.7 pc.
(Belian *et al.*, 1976c; Lategan, 1978).

17300 ← 333   X-ray: Rapid burster. States with rapid sequence of bursts are observed. Duration of bursts ~10–50 s, intervals between bursts from 15 s to 5–10 min, $kT > 10$ keV, $N_H \sim 3 \times 10^{22}$ cm^{-2}, $T \sim (1$–2$) \times 10^{7\circ}$.
Periods of activity are repeated every 6.5 months.
Optics: infrared observations give weak object, which may be globular cluster at the distance ~30 kpc.
Size of the emission region 10–20 km. Distance 9^{+5}_{-2} kpc.
(Calla *et al.*, 1978; Hoffman *et al.*, 1978; Hearn *et al.*, 1976a; Heise and Grindlay, 1976; Hoffman, 1976; Kleinmann *et al.*, 1976a; Kleinmann, 1976; Lewin and Joss, 1977; Lewin, 1976a, 1976b; Lewin *et al.*, 1976a, b, c, 1977a; Liller, 1976c; Marshall *et al.*, 1979; Marshall, 1978; Marshall *et al.*, 1978a; Ulmer *et al.*, 1977; van Paradijs *et al.*, 1978a; van Paradijs *et al.*, 1979; van Paradijs *et al.*, 1978; Watson *et al.*, 1976; White *et al.*, 1977).

17300 − 238   Many bursts of duration ~1 s.
(Lategan, 1978).

17309 − 220   X-ray: transient $\tau_{min.} > 60$ days, $\tau_e \sim 30$ days, $kT = 4 \pm 1$ keV.
(Cominsky *et al.*, 1978; Forman *et al.*, 1978).

17329 − 449   X-ray: Hard source. Burst of duration 5–15 s.
Possibly XRS 17353 − 444 (?).
(Sagdeev, 1976).

Table II (continued)

17342 − 127	NRAO 530. Radio source, bursts.
	(Marshall *et al.*, 1979).
17353 − 444	Steady source has $kT = 7$–15 keV, time of burst ~ 2 s.
	Optics: star $V = 17.5$, strong blue continuum. $B \approx 17.7$, distance ~ 5–10 kpc, $B - V = 0.22$; $U - B = -0.82$, $kT = 8 \pm 2$ keV.
	Optics: burst of duration ~ 9 s, $A_v \approx 0.5$.
	Optic. burst delay on 2.8 s. Temperature of black body $(25$–$30) \times 10^6$ K. Absence of norm. star line absorption, but there is emission lines $\lambda\lambda$ 4630–40 (N III) and λ 4686 (He II) Doppler velocity 1000 km s^{-1}. Optic. burst energy $E_{opt.} = (1.5 \pm 0.4) \times 10^{-12}$ erg s^{-2}.
	(Abramenko *et al.*, 1978; Bond, 1977; Canizares and McClintock, 1979; Forman *et al.*, 1978; Friedman *et al.*, 1967; Grindlay *et al.*, 1978; Grindlay *et al.*, 1978a; Hackwell *et al.*, 1979; Jones, 1976; McClintock *et al.*, 1979; McClintock *et al.*, 1978; McClintock, 1977; McClintock, 1977; McClintock *et al.*, 1977b; Parsignault and Grindlay, 1978; van Paradijs *et al.*, 1979).
17354 − 284	X-ray: transient $kT = 3$–7 keV, $E_a \geqslant 2.6$ keV, $\tau_{min.} = 52$ days, $T = 4.8$ keV.
	(Cominsky *et al.*, 1978; Jones, 1976; Kellog *et al.*, 1971; Markert *et al.*, 1978; Villa *et al.*, 1976).
17360 − 163	Many bursts of duration ≈ 1 s.
	(Belian *et al.*, 1976c).
17400 − 463	X-ray: many bursts $\gtrsim 1$ s.
	(Belian *et al.*, 1976c).
17401 − 391	λ Sco (?).
	X-ray: binary system with $P_{orb.} = 5.6$ days, $V = 1.62$, spectrum B1. Distance ~ 100 pc, then $L_x = 2 \times 10^{33}$ erg s^{-1} (0.4–2 keV). Flux ~ 130 μJy.
	(Blicker *et al.*, 1973).
17417 − 296	X-ray: bursts of duration 10 s, intervals ~ 0.55 days.
	(Clark, 1976; Lewin, 1976a; Lewin *et al.*, 1976d).
17424 − 289	X-ray: transient $kT = 6$ keV, $N_H = 6.2 \times 10^{22}$ cm^{-2} (24.2.75), $\tau_{min.} > 18$ days, $kT = 3 \pm 1$ keV.
	(Carpenter *et al.*, 1977; Cominsky *et al.*, 1978; Branduardi *et al.*, 1976; Davies *et al.*, 1976; Eyles *et al.*, 1975b; Proctor *et al.*, 1978; Willmore, 1977).
17426 − 292	X-ray: burst of duration 30–40 s (3–6 keV), intervals nearly 1.46 days. Possibly, from this region the optic. bursts were observed.
	(Clark, 1976; Byrne and Waimann, 1977; Markert *et al.*, 1978; Lewin, 1976a; Lewin *et al.*, 1976d; Lewin, 1977).
17428 − 294	Carpenter *et al.*, (1977), Glass (1979), Jernigan *et al.* (1978), Proctor *et al.* (1978).
17430 − 295	X-ray: transient.
	IAU Circ., No. 2934.
	(*Cominsky et al.*, 1978).
17436 − 291	X-ray: source in galactic centre.
	It may be steady state of some burst. sources.
	(Forman *et al.*, 1978; Giacconi *et al.*, 1965; Haymes *et al.*, 1969; Jones, 1976).
17436 − 285	X-ray: bursts with intervals ~ 10 min.
	(Lewin, 1977).
17437 − 316	X-ray: transient $E_a \sim 1$ keV, distance 3 kpc.
	(Griffiths *et al.*, 1978; Holt, 1977a; Kaluzienski, 1977a; Wood *et al.*, 1978).
17438 − 288	X-ray: $kT \sim 2.9$ keV, $N_H \sim 2.5 \times 10^{23}$ cm^{-2}.
	(Proctor *et al.*, 1978).
17441 − 284	X-ray: 3 bursts with intervals ~ 17 and 4 min.
	(Hoffman, 1976; Lewin *et al.*, 1976d).
17448 − 361	Transient (?).
	(Carpenter *et al.*, 1977; Cominsky *et al.*, 1978; Davison *et al.*, 1976).
17448 − 265	X-ray: $kT = 3$–7 keV, $E_a \leqslant 2$–6 keV.
	(Andrew *et al.*, 1970; Cruddace, 1971; Cruddace *et al.*, 1972; Dolan, 1970; Fisher *et al.*, 1968; Forman *et al.*, 1976, 1978; Gorenstein, 1969; Hill *et al.*, 1972; Jones *et al.*, 1972; Jones, 1976; Lewin, 1976; Lewin *et al.*, 1978a; Meyer *et al.*, 1970; Markert *et al.*, 1978; Parsignault and Grindlay, 1978; Ricker *et al.*, 1976a).
17453 + 390	Clark *et al.* (1975a), Cooke *et al.* (1978), Forman *et al.* (1978), Markert *et al.* (1976a).

Table II (continued)

17456+292   Forman *et al.* (1978).

17460−203   X-ray: transient $kT = 4$ keV, $E_a = 1.6$ keV, $\tau_{min.} > 45$ days, $kT = 4 \pm 0.5$ keV.
Optics: globular cluster NGC 6440; $V = 9.28$; spectrum G3, class V, distance 6 kpc, $\alpha = 17^h45^m54^s6, \delta = -20°20'36''$.
(Cominsky *et al.*, 1978; Forman *et al.*, 1976b, 1978; Jones, 1976; Markert *et al.*, 1975b; Markert *et al.*, 1978; Ulmer *et al.*, 1976).

17468−370   Globular cluster NGC 6441.
X-rays: $kT = 3$–7 keV, $E_a \leqslant 2.6$ keV. Bursts with intervals 0.643 days.
Optics: $V = 7.13$, spectrum G2, class III, distance 10–18 kpc. Total mass $2.5 \times 10^5 \, M_\odot$. $\alpha = 17^h46^m48^s8, \delta = -37°02'25''$.
(Bahcall, 1976; Clark *et al.*, 1975; Forman *et al.*, 1978; Gorenstein, 1969; Hill *et al.*, 1972; Jones, 1976; Jernigan *et al.*, 1978; Jernigan and Clark, 1979).

17477−293   Extended (0.4–1°) region or complex source.
(Proctor *et al.*, 1978).

17491−284   X-rays: $kT \sim 2.9$ keV, $N_H \sim 2.5 \times 10^{23}$ cm^{-2}.
(Proctor *et al.*, 1978).

17522−008   Marshall *et al.* (1979).

17536+151   X-rays: L_x (0.2–2.8 keV) $< 2 \times 10^{31}$ erg s^{-1}.
Optics: HD 163950, $d = 85$ pc, $m = 7.3$–8.1, $P = 3.992\,801\,2$ days, spectrum F4 IV − V + K0 IV.
(Walter *et al.*, 1978a).

17555−338   X-rays: $kT \leqslant 3$ keV, $E_a \leqslant 2.6$ keV.
(Fisher *et al.*, 1968; Forman *et al.*, 1978; Gorenstein, 1969; Hill *et al.*, 1972; Jernigan *et al.*, 1978; Jones, 1976; Markert *et al.*, 1978; Parsignault and Grindlay, 1978).

17585−205   X-rays: $kT = 3$–7 keV, $E_a \leqslant 2.6$ keV.
(Dolan, 1970; Fisher *et al.*, 1968; Forman *et al.*, 1978; Friedman *et al.*, 1967; Gorenstein and Giacconi, 1968; Gorenstein, 1969; Jones, 1976; Lewin *et al.*, 1968; Markert *et al.*, 1978; Meyer *et al.*, 1970; Parsignault and Grindlay, 1978; Rappaport *et al.*, 1969; Seward *et al.*, 1972; Seward, 1970; Willmore *et al.*, 1974).

17590−664   Forman *et al.* (1978).

17590−250   X-ray: $kT = 4.3$ keV.
(Andrew *et al.*, 1970; Braes *et al.*, 1972; Cruddace *et al.*, 1972; Davidsen *et al.*, 1976; Dolan, 1970; Fisher *et al.*, 1966, 1968; Forman *et al.*, 1978; Glass, 1979; Gorenstein and Giacconi, 1968; Jernigan *et al.*, 1978; Jones, 1976; Lewin *et al.*, 1968a; Markert *et al.*, 1968; Meyer *et al.*, 1970; Novikova *et al.*, 1977; Parsignault and Grindlay, 1978; Rappaport *et al.*, 1969; White *et al.*, 1975; Willmore *et al.*, 1974).

18004+682   Marshall *et al.* (1979).

18010+698   Object of BL Lac type 3C 371 (?).
X-ray: luminosity (2–6 keV) $\sim 7 \times 10^{43}$ erg s^{-1}.
Optics: $V = 14.8$, variability ~ 2.
(Schwartz *et al.*, 1978b; Marshall *et al.*, 1979).

18025−448   X-ray: burst of duration 5–10 s.
(Sagdeev, 1976).

18034−605   (Forman *et al.*, 1978).

18037−246   Cominsky *et al.* (1978), Jernigan (1976a), Jernigan *et al.* (1978).

18055−186   Watson *et al.* (1976).

18061+458   Nova star DQ Her.
X-ray: in the range 260–1200 keV, intensity 1.6×10^{-5} phot cm^{-2} keV^{-1}.
(Coe *et al.*, 1977).

18063−274   4 bursts of duration ~ 30 s and intervals 0.86–1.1 days.
(Swank *et al.*, 1976).

18070−365   Possibly, is source 18204 − 303 (?). Burst of duration less than 128 s.
(Evans *et al.*, 1976).

18079−108   Cominsky *et al.* (1977b, 1976), Forman *et al.* (1978).

18117−171   X-ray: spectrum $kT = 3$–7 keV, $E_a \leqslant 2.6$ keV, $N_H = 6 \times 10^{22}$ cm^{-2}.
Optics: $A_v \sim 15$, IR-region $K \sim 5.0$.

Table II (continued)

	(Bowyer *et al.*, 1965; Bradt *et al.*, 1971; Cruddace *et al.*, 1972; Davidsen *et al.*, 1976; Dolan, 1970; Gorenstein and Giacconi, 1968; Fisher *et al.*, 1966; Forman *et al.*, 1978; Gorenstein, 1969; Glass, 1979; Jones, 1976; Markert *et al.*, 1978; Meyer *et al.*, 1970; Parsignault and Grindlay, 1978; Reina, 1974; Rappaport *et al.*, 1969; Seward *et al.*, 1972; White *et al.*, 1978c).
18118 + 379	Forman *et al.* (1978).
18124 − 121	Forman *et al.* (1978), Markert *et al.* (1978).
18131 − 140	X-ray: pulsar P = 31.88 min., spectrum kT = 4–10 keV, $N_{\rm H} \approx (1–8) \times 10^{22}$ cm^{-2}. Optics: star V = 17.51, $B − V$ = 1.26; $U − B$ = 1.03; period 31.88 min. Total mass of system $\sim 2\,M_\odot$.
	(Andrew *et al.*, 1970; Bradt *et al.*, 1971; Cruddace, 1971; Cruddace *et al.*, 1972; Davidsen *et al.*, 1976; Dolan, 1970; Fisher *et al.*, 1966, 1968; Forman *et al.*, 1976c, 1978; Friedman *et al.*, 1967; Fujimoto *et al.*, 1969; Gorenstein *et al.*, 1968; Kasturirangan *et al.*, 1976; Markert *et al.*, 1978; Margon, 1978; Meyer *et al.*, 1970; Parsignault, 1978; Reina and Grindlay, 1974; Seward, 1970; Seward *et al.*, 1972; White *et al.*, 1976b; White *et al.*, 1978c).
18149 + 498	X-ray: binary system, $P_{\rm orb.}$ = 0.128 927 days, eclipse 0.15 of period, spectrum $kT \approx$ 18 keV (38–80 keV), kT = 0.025 keV (0.15–0.5 keV). In the range 0.15–0.5 keV the intensity 8 × 10^{-10} erg cm^{-2} s^{-1}, polarization; luminosity (>2 keV) $\sim 4 \times 10^{32}$ erg s^{-1}. Optics: AM Her V = 12.20; $B − V$ = 0.58, $U − B$ = 0.05, K = 284 km s^{-1}. System red and magnetic white dwarfs.
	(Bunner, 1978; Cowley *et al.*, 1976a; Cooke *et al.*, 1978; Hear *et al.*, 1976a; Murray and Ulmer, 1976; Markert *et al.*, 1978; Olson, 1977; Priedhorsky, 1977; Szkody and Brownlee, 1977; Topka, 1977; Tuohy *et al.*, 1978a).
18159 − 083	Villa *et al.* (1976).
18162 − 123	Villa *et al.* (1976).
18175 − 059	Possibly, is source 18351 − 078 (?).
	(Forman *et al.*, 1978; Seward *et al.*, 1976b).
18204 − 303	X-ray: steady source and bursts. Duration of bursts ~ 10 s, intervals 0.12–0.18 days, spectrum $kT \approx 5.5$ keV. Optics: globular cluster NGC 6624, V = 8.24, spectrum G5, A_v = 1.3, distance 5–8 kpc, mass 1.6 × 10$^5\,M_\odot$.
	(Canizares *et al.*, 1978; Clark *et al.*, 1975a, 1976; Bahcall, 1976; Forman *et al.*, 1978; Grindlay *et al.*, 1977a; Grindlay, 1978; Jones, 1976; Jernigan and Clark, 1979; Markert *et al.*, 1978; Parsignault and Grindlay, 1978; van Paradijs *et al.*, 1979).
18217 − 054	X-ray: 11 bursts of duration 5–10 s and intervals 0.16–0.25 days.
	(Swank *et al.*, 1976c).
18223 − 371	X-ray: $N_{\rm H} \sim 8 \times 10^{22}$ cm^{-2}, power spectrum, index -0.7. Optics: star V = 16.29, $B − V$ = −0.05, emission C III/N III, He II 4686.
	(Cooke *et al.*, 1978; Dolan, 1970; Forman *et al.*, 1978; Friedman *et al.*, 1967; Gorenstein, 1969; Griffiths *et al.*, 1978a; Markert *et al.*, 1978; Meyer *et al.*, 1970; Seward, 1970).
18228 − 000	X-ray: spectrum kT = 3–7 keV, $E_a \leqslant 2.6$ keV.
	(Doxey *et al.*, 1977; Forman *et al.*, 1978; Jones, 1976; Markert *et al.*, 1978; Seward *et al.*, 1976b).
18248 + 644	Active galaxy 3C 383 (?).
	(Marshall *et al.*, 1979).
18251 + 339	Source, possibly, confused.
	(Forman *et al.*, 1978).
18288 − 065	Villa *et al.* (1976).
18301 + 345	Source, possibly, confused.
	(Forman *et al.*, 1978).
18310 − 109	Villa *et al.* (1976).
18317 − 232	Forman *et al.* (1978), Markert *et al.* (1978).
18325 − 051	Forman *et al.* (1978), Markert *et al.* (1978), Seward *et al.* (1976a).
18332 + 326	N-galaxy 3G 382 (?). X-rays: luminosity (2–10) $\sim 5.5 \times 10^{44}$ erg s^{-1} (?). Optics: V = 14.7.
	(Marshall *et al.*, 1979; Marshall *et al.*, 1978).

Table II (continued)

18334−077	Marshall *et al.* (1979).
18341−626	Seyfert galaxy ESO-140-G43 (?). (Marshall *et al.*, 1979).
18347−653	Seyfert galaxy ESO 103-G 35 (?). (Marshall *et al.*, 1979).
18351−078	Transient (?). Possibly, is steady source with long-term variability. (Hill *et al.*, 1974).
18353+387	X-rays: luminosity (0.15–0.8 keV) $\sim 3 \times 10^{28}$ erg s^{-1}, emission measure $\sim 10^{51}$ cm^{-3}. Optics: Vega (α Lyr), $V \sim 0.04$, $B - V = 0$, $U - B = 0$, $m_v = 0.5$, spectrum AOV distance ~ 8.1 pc. (Topka *et al.*, 1979).
18358−114	Galactic cluster NGC 6649 (?). (Forman *et al.*, 1978; Seward *et al.*, 1978).
18360−227	2 bursts of duration ~ 15 s. (Bunner *et al.*, 1976a).
18374+049	X-rays: steady and burst source. Duration of bursts ~ 10 s, intervals ~ 6.3 h, spectrum $kT = 3$–7 keV (steady source), $E_a \leqslant 2.6$ keV. Optics: star $V = 17.65$; $B = 18.68$; 7 September, 1978, burst (at the same time with X-rays). (Abramenko *et al.*, 1978; Bowyer *et al.*, 1965; Brini *et al.*, 1967; Bernacca *et al.*, 1979; Davidsen *et al.*, 1976; Dolan, 1970; Fisher *et al.*, 1968; Forman *et al.*, 1976c, 1978; Friedman *et al.*, 1967; Gorenstein, 1969; Hackwell *et al.*, 1979; Jones, 1976; Lewin *et al.*, 1968; Li *et al.*, 1977; Markert *et al.*, 1978; Parsignault and Grindlay, 1978; Reina, 1974; Seward, 1970; Seward *et al.*, 1972, 1976b; Torstensen *et al.*, 1978a; Ulmer *et al.*, 1978; van Paradijs *et al.*, 1979).
18383+629	Marshall *et al.* (1979).
18389+378	X-ray: the most intensity 0.0139 ± 0.0027 keV cm^{-2} s^{-1}. Spectrum $kT \lesssim 0.4$ keV. Optics: dwarf Nova AY Lyr (?). (Cordova and Garmire, 1978).
18400+013	Seward *et al.* (1976b), Villa *et al.* (1976).
18434+675	Markert *et al.* (1978).
18456−024	Bradt *et al.* (1978), Doxey *et al.* (1977), Forman *et al.* (1978), Seward *et al.* (1976b).
18476−053	Galactic cluster NGC 6704 (?). (Seward *et al.*, 1976b).
18476+789	N-galaxy 3C 390.3 (?). X-ray: luminosity (2–10 keV) $\sim 2.4 \times 10^{44}$ erg s^{-1}. Optics: $V = 14.5$. (Forman *et al.*, 1978; Marshall *et al.*, 1978; Marshall *et al.*, 1979).
18482−079	Burst of duration 10–20 s. (Swank *et al.*, 1976c).
18490−771	Forman *et al.* (1978).
18492−312	Forman *et al.* (1978).
18503−087	X-ray: burst of duration ~ 78 min. (spectrum $kT = 4.5$ keV, $E_a < 2.3$ keV). Luminosity of steady source (2–6 keV) $\sim 5 \times 10^{37}$ erg s^{-1}. Optics: globular cluster NGC 6712, $V = 8.13$, spectrum G0.3, distance 5.7 kpc, radius of nucleus 1.6 pc. (Cominsky *et al.*, 1977; Bradt *et al.*, 1978; Doxey *et al.*, 1977; Forman *et al.*, 1978; Grindlay *et al.*, 1977a).
18508+007	Seward *et al.* (1976b).
18526+370	Forman *et al.* (1978).
18536−236	Forman *et al.* (1978).
18536+012	Supernova remnant W 44 (?). X-ray: luminosity (1–3.5 keV) is equal to $(0.5–5.4) \times 10^{35}$ erg s^{-1}, spectrum $kT = 0.1$–3.3 keV, $N_H = (0.3–10.0) \times 10^{21}$ cm^{-2}. Optics: distance 3 kpc, diam 24 pc.

Table II (continued)

	Radio: age 3–14 thousand yrs, $E_a = (5$–$35) \times 10^{50}$ erg.
	(Gronenschild *et al.*, 1978; Winkler, 1978).
18543+683	Cluster of galaxies Abell 2312 (?).
	X-ray: in the range 2–10 keV, luminosity 6×10^{44} erg s^{-1}.
	(Cooke *et al.*, 1978; Forman *et al.*, 1978; McHardy, 1978).
19010+430	X-ray: in the region 0.18–0.43 keV intensity 4×10^{-12} erg cm^{-2} s^{-1}, luminosity $(4$–$60) \times 10^{31}$
	erg s^{-1}, spectrum $0.007 < kT < 0.86$ keV.
	Optics: system MVLyr, type UX UMa, orbital period 2 h, distance 300–1200 pc.
	(Mason *et al.*, 1979).
19017+031	Transient source.
	X-ray: spectrum $kT > 15$ keV, $E_a < 2.6$ keV, life-time 80 days.
	(Cominsky *et al.*, 1978; Forman *et al.*, 1978; Jones, 1976; Seward *et al.*, 1976a; Villa *et al.*, 1976).
19048+670	Markert *et al.* (1978).
19059+000	X-ray: steady and burst. states.
	Burst with duration of several seconds.
	(Bradt *et al.*, 1978; Doxey *et al.*, 1977; Forman *et al.*, 1978; Lewin *et al.*, 1976c; Lewin, 1976b;
	Seward *et al.*, 1976b; van Paradijs *et al.*, 1979; Villa *et al.*, 1976).
19078+095	Forman *et al.* (1978), Markert *et al.* (1978), Seward *et al.* (1976).
19082+023	X-ray: one burst.
	(Li and Lewin, 1976).
19087+005	X-ray: transient source with steady state.
	Transient phenomena of duration ~ 1 month, spectrum $kT = 3.5$ keV.
	Optics: in steady state star $V = 19.17; B - V = 1.26; M_v = 5.9$, during burst $V \sim 17$, optic. burst
	correlates with X-ray burst, activity $\Delta V = 0.2$ with characteristic time 1–2 min, weak emission
	C III-N III, He II. Distance 2.5 kpc.
	(Charles *et al.*, 1978c; Cominsky *et al.*, 1978; Davidsen *et al.*, 1976; Doxey *et al.*, 1977; Forman *et al.*,
	1978; Friedman *et al.*, 1967; Holt and Kaluzienski, 1977; Jones, 1976; Kaluzienski and Holt,
	1978c; Markert *et al.*, 1978; Novikova *et al.*, 1977; Seward, 1970; Seward *et al.*, 1976a; Schwartz *et*
	al., 1972; Thorstensen *et al.*, 1978; Walter *et al.*, 1978; Watson, 1976).
19092+076	Forman *et al.* (1978), Markert *et al.* (1978), Seward *et al.* (1976b).
19094+047	X-ray: there is no periodic variations from 0.16 to 5 s.
	Optics: emission object SS 433, $V \sim 13$, $M_v \leqslant -3.5$, distance ~ 2.5–3.5 kpc, series of emission
	lines with confusion up to $+55\,000$ km s^{-1}, $-30\,000$ km s^{-1}, period 164 ± 3 days, steady lines
	with $K \sim 100$ km s^{-1}, $P = 13.2$ days.
	Radio: supernova remnant W 50 (extended object) and point source, nonthermal spectrum.
	(Forman *et al.*, 1978; Marshall *et al.*, 1978b; Margon, 1978a; Seward *et al.*, 1976b; Seaquist *et al.*,
	1979).
19117−108	Transient (?).
	(Hill *et al.*, 1974).
19146−589	Seyfert galaxy ESO 141-G 55 (?).
	X-ray: luminosity (2–10 keV) $\sim 2 \times 10^{44}$ erg s^{-1}.
	(Cooke *et al.*, 1978; Forman *et al.*, 1978; Griffiths *et al.*, 1979).
19160−793	Cluster of galaxies SC 1834−772 (?).
	X-ray: luminosity (2–10 keV) $\sim 6.4 \times 10^{43}$ erg s^{-1}.
	Radio: source PKS 1833−77.
	(Cooke *et al.*, 1978; Forman *et al.*, 1978; Marshall *et al.*, 1979; Lugger, 1978).
19161−053	X-ray: steady and burst source, 18 bursts with intervals 4–5 h, spectrum $kT \sim 35$ keV.
	(Doxey *et al.*, 1977; Forman *et al.*, 1978; Joss, 1977; Lewin and Joss, 1977; Lewin and Hoffman,
	1977; Markert *et al.*, 1978; Price *et al.*, 1972; Takagishi *et al.*, 1978; van Paradijs *et al.*, 1979).
19185+146	X-ray: transient source observed during 2 weeks (Uhuru) spectrum $kT \approx 6$ keV.
	(Cominsky *et al.*, 1977a; Cominsky *et al.*, 1978; Forman *et al.*, 1978; Gursky *et al.*, 1967; Villa *et al.*,
	1976).
19197+436	Cluster of galaxies Abell 2319 (?).
	X-ray: luminosity (2–10 keV) $= 1.6 \times 10^{42}$ erg s^{-1}, spectrum $kT = 12.5$ keV.
	(Cooke *et al.*, 1978; Forman *et al.*, 1978; Grindlay *et al.*, 1977b; Henry and Tucker, 1979; Markert

Table II (continued)

	et al., 1978; Mushotzky *et al.*, 1978; McHardy, 1978; Rowan-Robinson and Fabian, 1975; Schwartz *et al.*, 1979a).
19202 + 340	Forman *et al.* (1978).
19262 + 506	Marshall *et al.* (1979).
19290 + 079	Bursts of duration ~ 10 s. (Clark, 1976).
19336 + 361	Forman *et al.* (1978).
19365 + 326	Supernova remnant (?). X-ray: in the range 0.5–2.0 keV intensity 2.4×10^{-10} erg cm^{-2} s^{-1}, spectrum $kT = 0.08$–0.8 keV, $N_{\mathrm{H}} = (2.5$–$260) \times 10^{20}$ cm^{-2}. Radio: angle diam $3\overset{\circ}{.}5$, age $(1.5$–$4.6) \times 10^4$ yr. (at the distance 70 pc), $E_a = 4 \times 10^{50}$ erg. (Snyder *et al.*, 1978).
19386 − 105	Seyfert galaxy NGC 6814 (?). (Cooke *et al.*, 1978).
19434 + 364	Forman *et al.* (1978).
19490 + 440	X-ray: burst of duration 18 s. (Nishimura *et al.*, 1978).
19540 + 319	X-ray: spectrum $kT = 7$–15 keV, $E_a \leqslant 2.6$ keV. (Forman *et al.*, 1978; Jones, 1976; Markert *et al.*, 1978).
19556 − 698	Cluster of galaxies CA 1955 − 692 or CA 2013 − 710 (?). X-ray: luminosity (2–10 keV) $\sim 2.5 \times 10^{45}$ erg s^{-1} (at identification with CA 1955 − 692), or 2.5×10^{43} erg s^{-1} (at identification with CO 2013 − 710). (Cooke *et al.*, 1978; Forman *et al.*, 1978; Markert *et al.*, 1976a; Lugger, 1978; Melnick and Quintana, 1975).
19560 + 650	Markert *et al.* (1978).
19564 + 350	Binary system HDE 226868. X-ray: orbital period 5.6 days. Fluctuations of intensity with duration of milliseconds, luminosity (2–10 keV) $\sim 1.5 \times 10^{37}$ erg s^{-1}, spectrum $kT > 15$ keV and $kT < 3$ keV. Optics: supergiant, $B = 9.8$, spectrum O9.7 Iab, $K = 72$ km s^{-1} distance 2.5 kpc, 6–7.75 optical pulsations with period 83 ms, period 39 (or 78 days) is found. Mases of components 25 $M_{\odot}$ (supergiant) and 7–14 $M_{\odot}$ (Black hole ?). Radio: weak source, flux 0.01–0.02 f.u. (Auriemma *et al.*, 1976; Aksenov, 1976; Berezhnoy *et al.*, 1977; Bolton, 1976; Canizares and Oda, 1977; Chi-Chao *et al.*, 1976a; Forman *et al.*, 1978; Frontera and Fuligni, 1975, 1976; Holt *et al.*, 1976b; Hutchings, 1974; Jain *et al.*, 1976; Jones, 1976; Kaluzienski *et al.*, 1976a; Kemp, 1977; Kemp *et al.*, 1977, 1978, 1978a; Liller, 1976b; Matteson *et al.*, 1976; Milgrom, 1978; Markert *et al.*, 1978; Natali and Messi, 1978; Rothschild *et al.*, 1976).
19570 + 115	X-ray: luminosity (2–10 keV) $\sim 7 \times 10^{36}$ erg s^{-1} (?). Optics: star $V = 18.7$, $B - V = 0.3$, $U - B = -0.6$, $A_v \sim 1.0$. Distance ~ 7 kpc. (Bradt *et al.*, 1978; Doxey *et al.*, 1977; Forman *et al.*, 1978; Markert *et al.*, 1978; Seward *et al.*, 1976b).
19572 + 405	Radio: galaxy Cyg A. X-ray: luminosity (2–6 keV) $\sim 4.7 \times 10^{44}$ erg s^{-1}, spectrum $kT \sim 6.5$ keV. (Forman *et al.*, 1978; Fabbiano *et al.*, 1979; Henry and Tucker, 1979; Markert *et al.*, 1978).
20014 + 626	Cooke *et al.* (1978), Forman *et al.* (1978).
20036 + 643	Forman *et al.* (1978).
20091 − 569	Cluster of galaxies (?). (Cooke *et al.*, 1978; McHardy, 1978).
20171 + 368	X-ray: spectral index -0.8 ± 0.1. Association with gamma-source CG 075 + 0 is possible. (Julien and Helmken, 1978).
20190 + 395	X-ray: spectral index -0.8 ± 0.1. Association with gamma-source CG 078 + 1 is possible. (Forman *et al.*, 1978; Julien and Helmken, 1978).

Table II (continued)

20200+405	Supernova remnant DR 4 (?).

20200+405   Supernova remnant DR 4 (?).
X-ray: luminosity (0.5–2 keV) is equal to (0.3–1) $\times$ 10^{35} erg s^{-1}, spectrum $kT = 0.1$–0.3 keV, $N_{\mathrm{H}} = 10^{22}$ cm^{-1}.
Optics: distance 2 kpc, diam 30 pc (?).
Radio: source DR 4.
(Davidsen *et al.*, 1977; Winkler, 1978a).

20288+428   Forman *et al.* (1978).

20305+407   X-ray: binary system. $P_{\mathrm{orb.}} = 0.199\,679\,47$ days, period 16.9 days may also exist. Steady source and burst. Spectrum $kT = 3$ keV, $N_{\mathrm{H}} = 5 \times 10^{22}$ cm^{-2}, luminosity (2–6 keV) $\sim 2 \times 10^{38}$ erg s^{-1}. IR-region; $H = 12.5$; $K = 11.4$, period 0.2 days is observing, $A_x = 1.5$, $A_v \sim 19$, distance ~ 15 kpc.
Gamma-source: CG 75+0 (?) or CG 78+1 (?).
Radio: variable source, bursts up to 21 f.u. (10 GHz).
(Apparao, 1977; Boldt *et al.*, 1976b; Becker *et al.*, 1978a; Dolan, 1970; Forman *et al.*, 1976c, 1978; Friedman *et al.*, 1967; Gainer *et al.*, 1976; Gorenstein *et al.*, 1968; Jones, 1976; Kestenbaum *et al.*, 1978; Lamb *et al.*, 1979; Mason *et al.*, 1976a; Milgrom, 1976; Margert *et al.*, 1978; Mando *et al.*, 1978; Marashi *et al.*, 1977; Parsignault *et al.*, 1976b, 1976c; Pietsch *et al.*, 1976; Ricker *et al.*, 1976b; Seward, 1970; Seward *et al.*, 1972).

20312+317   Villa *et al.* (1976).

20332+362   Villa *et al.* (1976).

20380+290   Soft source, after 1967 was not observed.
(Friedman *et al.*, 1967; Henry *et al.*, 1968; Meekins *et al.*, 1969).

20407−115   Seyfert galaxy MKN 509 (?).
X-ray: luminosity (2–10 keV) $\sim 2.5 \times 10^{44}$ erg s^{-1}.
Optics: $V = 13.12$, distance 210 mpc.
IR-region: on 1.25 μm intensity 0.028 f.u.
(Bradt, 1979; Cooke *et al.*, 1978).

20419+ 754   Markert *et al.* (1978).

20467+319   Possibly, is source 20497+308 (?).
(Forman *et al.*, 1978).

20486+443   Forman *et al.* (1978).

20497+308   Supernova remnant Cyg Loop.
X-ray: in the range 0.5–2 keV, intensity 9 $\times$ 10^{-9} erg cm^{-2} s^{-1}, luminosity $\sim 6 \times 10^{35}$ erg s^{-1} (0.5–2 keV), spectrum $kT = 0.2$ keV, $N_{\mathrm{H}} = 5.5 \times 10^{20}$ cm^{-2}. In the rocket flight of 30.3.73 hot point in Cygnus Loop observed (0.15–1.8 keV), spectrum $kT = 0.4$ keV.
Observed pulsation with period 62 ms.
Optics: age 1.7 $\times$ 10^4 yr., distance 0.8 kpc.
(Bleach *et al.*, 1975; Charles *et al.*, 1975a; Culhane, 1977; Gorenstein *et al.*, 1974; Rappaport *et al.*, 1974a; Winkler, 1978a).

20560+493   Forman *et al.* (1976a, 1978), Markert *et al.* (1978).

20582+328   Forman *et al.* (1978).

20586+417   Marshall *et al.* (1979).

21040+315   Forman *et al.* (1978).

21091+385   Transient source (?).
(Friedman *et al.*, 1967; Seward, 1970).

21200+450   X-ray: soft extended source, spectrum $kT = 0.06$ keV, $N_{\mathrm{H}} = 3 \times 10^{22}$ cm^{-2}.
(Coleman *et al.*, 1971; Davidsen *et al.*, 1977).

21203+321   Forman *et al.* (1978).

21275+119   Globular cluster M15 (NGC 7078).
X-ray: luminosity (2–6 keV) from 0.8 to 4 $\times$ 10^{36} erg s^{-1}.
Optics: $V = 6.29$, spectrum F9, distance 10.5 min.
(Bahcall, 1976; Bradt, 1978; Cooke *et al.*, 1978; Clark *et al.*, 1975a; Clark, 1979; Forman *et al.*, 1978; Jernigan and Clark, 1979; Markert *et al.*, 1978; Mironov and Scherepaschuk, 1976; Markert *et al.*, 1978; Newell *et al.*, 1976; Ulmer *et al.*, 1976).

21288+816   Markert *et al.* (1978).

Table II (continued)

21296+471	X-ray: luminosity (3–10 keV) $\sim 6 \times 10^{34}$ erg s^{-1}.
	Optics: star $V = 16.2$–17.4, $A_v \geqslant 0.7$, emission He II, C III/N III, binary system, $P_{orb.} = 0.2186$ days.
	(Bradt *et al.*, 1978; Forman *et al.*, 1978; Markert *et al.*, 1978; Thorstensen, 1979).
21346+557	Forman *et al.* (1978).
21370+567	X-ray: transient source, lifetime $\geqslant 15$ days, spectrum $kT \geqslant 15$ keV.
	(Cominsky *et al.*, 1978; Ulmer *et al.*, 1973).
21407+433	X-ray: soft source, in range 0.15–0.28 keV, intensity 5×10^{-10} erg cm^{-2} s^{-1}. During optical bursts soft (0.16–0.28 keV) and hard (1–7 keV) fluxes were observed, in the range 0.15–0.28 keV, luminosity 1.5×10^{33} erg s^{-1}.
	Binary system, $P_{orb.} = 0.128\,92$ days. During bursts period of 0.9 s (0.2–0.4 keV) was observed.
	Optics: dwarf Nova SS Cyg.
	$V = 12.1$–8.1, bursts with intervals 33–71 days, $M = 7.0$–3.0, distance ~ 100 pc, binary system $P_{orb.} = 6.63$ h, during bursts modulation with period 9.735 s.
	(Bradt *et al.*, 1978; Cordova *et al.*, 1978; Cordova, 1978; Fabbiano *et al.*, 1978a; Heise *et al.*, 1978; Mason *et al.*, 1978b; Patterson *et al.*, 1978; Rappaport *et al.*, 1974b; Watson *et al.*, 1978a).
21406−602	Cluster of galaxies SC 2146−594 (?).
	X-ray: luminosity (2–6 keV) $\sim 4.3 \times 10^{44}$ erg s^{-1}.
	(Forman *et al.*, 1978; Lugger, 1978; Markert *et al.*, 1976a; Markert *et al.*, 1978; Murray and Ulmer, 1976).
21426+380	X-ray: binary system, period 9.8 days, spectrum $kT = 3$–7 keV, $E_a \leqslant 2.6$ keV, luminosity (2–6 keV) $\sim 4 \times 10^{36}$ erg s^{-1}.
	Optics: star $V = 1341$ Cyg, $V = 14.7$, binary system, $P_{orb.} = 9.84$ days (Period 0.8614 days also measured), distance 2.1 kpc, $A_v \sim 1.0$.
	(Bradt *et al.*, 1978; Basco *et al.*, 1976; Chevalier *et al.*, 1976a; Crampton and Cowley, 1976; Crampton and Cowley, 1978; Forman *et al.*, 1978; Ilovaisky *et al.*, 1978c; Markert *et al.*, 1978; Novikova *et al.*, 1977; Parsignault and Grindlay, 1978).
21511+309	Markert *et al.* (1978).
21518−316	Cooke *et al.* (1978), Marshall *et al.* (1979).
21541−303	Probably, is source 21559−304.
	(Marshall *et al.*, 1979).
21554−609	Cooke *et al.* (1978).
21559−304	Object type BL Lac (?).
	X-ray: spectrum $kT = 0.3$ keV, luminosity (2–10 keV) $\sim 1.7 \times 10^{46}$ erg s^{-1}.
	Optics: $B = 14.5$; $B - V = 0.1$.
	IR-region, $J = 11.78$; $H = 11.23$.
	Radio: on frequency 2700 mHz flux 0.32 f.u.
	(Agrawal and Riegler, 1978; Grinstein *et al.*, 1978; Griffiths *et al.*, 1979c; Schwartz *et al.*, 1978b).
22046+452	X-ray: in the region 0.2–2.8 keV luminosity 1.5×10^{31} erg s^{-1}.
	Optics: AR Lac, system of RS CVn type, $V = 6.87$–7.69, orbital period $\sim 1.983\,216$ days, spectrum G5 IV + K0 IV, distance 50 pc.
	(Woltjer *et al.*, 1978).
22049+472	X-ray: in the range 0.2–2.8 keV, luminosity 1.3×10^{32} erg s^{-1}.
	Optics: HK Lac, system of RS CVn-type, period 24.428 days. $V = 6.52$, spectrum K0 III + F IV, distance 150 pc.
	(Walter *et al.*, 1978).
22063+544	Forman *et al.* (1978), Markert *et al.* (1978), Ulmer *et al.* (1973), Villa *et al.* (1976).
22092+261	Forman *et al.* (1978).
22095−471	Seyfert galaxy NGC 7213 (?).
	(Marshall *et al.*, 1979).
22116+124	Optics: star RU Peg., $V = 13.1$–10.0, binary system $P_{orb.} = 8.9$ h.
	(Watson *et al.*, 1978a).
22139+239	Forman *et al.* (1978).
22154−086	Marshall *et al.* (1979).

Table II (continued)

22204−022	Cluster of galaxies Abell 2440 (?). X-ray: luminosity (2–10 keV) $\sim$ 4.5 × 10^{44} ergs^{-1}. (Cooke *et al.*, 1978; Marshall *et al.*, 1979; McHardy, 1978; Schwartz, 1978).
22248−782	Forman *et al.* (1978).
22264+014	Cluster of galaxies Abell 2457 (?). (Marshall *et al.*, 1979).
22330−378	Marshall *et al.* (1979).
22373−256	Possibly, is source 22444−242 (?). (Cooke *et al.*, 1978).
22389+607	Forman *et al.* (1978), Markert *et al.* (1978).
22404+267	Forman *et al.* (1978).
22444−242	Markert *et al.* (1975a, 1976a, 1978), Murray and Ulmer (1976).
22468+601	Villa *et al.* (1976).
22514−178	Quasar MR 2251−178 (?). X-ray: luminosity (2–10 keV) $\sim$ (0.5–1.5) × 10^{45} ergs^{-1}, spectrum hard, index −0.5 ± 0.5. Optics: $V = 14.11$, $A_c = 0.31$, distance 385 mpc. (Bradt, 1979; Canisares *et al.*, 1978a; Cooke *et al.*, 1978; Ricker *et al.*, 1977; Ricker *et al.*, 1977a).
22525+181	Forman *et al.* (1978).
22528−035	Marshall *et al.* (1979).
22591+161	Forman *et al.* (1978), Ricketts (1978).
22595+085	Cooke *et al.* (1978), Forman *et al.* (1978).
23022−088	Cooke *et al.* (1978), Forman *et al.* (1978).
23086+597	Wood *et al.* (1978).
23120−214	Cluster of galaxies Abell 2554 (?). (Ricketts, 1978).
23153−428	Emission galaxy NGC 7582 (?). X-rays: luminosity 7 × 10^{42} ergs^{-1} (2–18 keV). Optics: $V = 13.5$, $A_v = 4.0$. (Cooke *et al.*, 1978; Ward *et al.*, 1978).
23166+618	Forman *et al.* (1978).
23184−272	Possibly, it is XRS 23210−230 (?). (Cooke *et al.*, 1978).
23210−230	Markert *et al.* (1975b, 1976a, 1978), Murray and Ulmer (1976).
23212+585	Supernova Remnant Cas A (1700?). X-rays: $L_x \approx 8 \times 10^{35}$ ergs^{-1} (2–6 keV), spectrum $kT \sim$ 1–4 keV, $N_H \simeq 4.3 \times 10^{21}$ cm^{-2}. Radio: distance 2.8 kpc, diam 3.5 pc, flux 2800 Jy (1 GHz), spectral index −0.76. (Culhane, 1977; Gorenstein *et al.*, 1974; Gorenstein and Tucker, 1975; Markert *et al.*, 1978; Walter, 1972; Winkler, 1978a).
23221+166	Cluster of galaxies Abell 2589 (?). X-rays: $L_x \sim 10^{44}$ ergs^{-1} (2–6 keV), $kT = 9$ keV, $N_H < 1.4 \times 10^{22}$ cm^{-2}. (Cooke *et al.*, 1978; Forman *et al.*, 1978, 1978a; McHardy, 1978; Muzchotsky *et al.*, 1978; Schwartz, 1978).
23357+427	Forman *et al.* (1978).
23440+086	Cluster of galaxies Abell 2657 (?). X-rays: luminosity (2–6 keV) 10^{44} ergs^{-1}. (Forman *et al.*, 1978; Forman *et al.*, 1978a; Ricketts, 1978; Schwartz, 1978).
23448−285	Cluster of galaxies K 44 (?). X-rays: luminosity (2–6 keV) $\sim$ 10^{44} ergs^{-1}. (Cooke *et al.*, 1978; Forman *et al.*, 1978; McHardy, 1978).
23454+273	Cluster of galaxies Abell 2666 (?). X-rays: luminosity (2–10 keV) $\sim$ 1.9 × 10^{44} ergs^{-1}. (Cooke *et al.*, 1978; Forman *et al.*, 1978; Forman *et al.*, 1978a; Markert *et al.*, 1978; Ricketts, 1978).
23512+067	Seyfert galaxy MKN 541 (?). X-rays: luminosity (2–10 keV) $\sim$ 2 × 10^{44} ergs^{-1}. (Cash *et al.*, 1979; Forman *et al.*, 1978; Ricketts, 1978; Schwartz, 1978).

Table II (continued)

23587+210	X-rays: burst of duration ∼ 1000 s.
	Steady flux also is observed.
	(Forman et al., 1978).
23588−640	X-rays: burst of duration less than 2 h, spectrum $kT \geqslant 20$ keV.
	(Cooke et al., 1976; Rappaport, 1975; Rappaport et al., 1976b; Willmore, 1975).
B-1	X-rays: 11 bursts with energies 10^{-6} erg cm^{-2}.
	(Meyer, 1976).
B-2	X-rays: burst of duration 5 h.
	(Cooke, 1976).
B-3	X-rays: burst of duration 5 h.
	(Cooke, 1976).
B-4	X-rays: burst in the range 0.2–1 keV intensity 1.7×10^{-9} erg cm^{-2} s^{-1}, luminosity 6.8×10^{31} erg s^{-1} (?).
	Optics: burst. star of type dM (?), distance 17.5 pc.
	(Lategan, 1978).

TABLE III

Galactic X-ray sources associated with binary systems

Source 2 XRS	kT (keV)	Optical spectrum	$M_{opt.}$	$P_{orb.}$ (days)	Distance (kpc)	Z (kpc)	L_x erg s^{-1} 2–6 keV	Comments
Massive systems								
01147 +650	?	B0.5 IIIe	20		1.5	0.7	2×10^{34}	steady
01152 +634	>15	Be	12–15	24	∼5	0.1	3×10^{37}	transient
03522 +308	≳10	O9.5 III–Ve	20	580?	0.35	0.1	2×10^{34}	steady
05357 +262	∼18	Bpe	12–25	20	1–3	0.1	2×10^{37}	transient
06560 −071	∼15	Be?	12–15?		2–5	0.1	2×10^{36}	transient
09002 −403	>15	B0.5 Ib	30	8.97	1.4	0.1	1.4×10^{36}	steady
11189 −615	>15	Be	12–15		3	0.04	2×10^{36}	transient
11190 −603	>15	O6.5 II–III	17	2.09	∼8	0.04	4×10^{37}	steady
11455 −619	7–15	B1 Vne	15		1.5	0.01	6×10^{36}	steady

Table II (continued)

Source 2 XRS	kT (keV)	Optical spectrum	$M_{opt.}$	$P_{orb.}$ (days)	Distance (kpc)	Z (kpc)	L_x erg s^{-1} 2–6 keV	Comments
12238 −624	>15	B1.5 Ia	40	22.6	2	0	10^{37}	steady
12582 −613	7–15	B6–9 pe	6–10		2	0.04	2×10^{36}	steady
15168 −569	>15	OI	∼40	16?	≳10	0	≳2×10^{38}	steady
15382 −521	>18	B0 I	20	3.7	∼7	0.25	4×10^{36}	steady
16531 −407	∼11	B0 Iae	20	7.8	2	0.05	4×10^{35}	steady
17005 −377	>15	O6.5 f	27	3.4	1.7	0.07	3×10^{36}	steady
19564 +350	≳15	O9.7 Iab	25	5.6	2.5	0.14	2×10^{37}	steady

Low mass systems

Source 2 XRS	kT (keV)	Optical spectrum	$M_{opt.}$	$P_{orb.}$ (days)	Distance (kpc)	Z (kpc)	L_x erg s^{-1} 2–6 keV	Comments
00421 +327	?	dwarf?	≲1	11.6?	≲5?	≲2.5?	≲2×10^{36}	bursts
04490 −550	?	M?	≲1		∼1	0.6	2×10^{35}	transient
06143 +091	?	emission	1–2		4–8	0.25–0.5	10^{37}	steady
06201 −003	1.4	G?	∼1	7.8	1–2	0.10–0.2	3×10^{38}	transient
09214 −630	?	emission	1–2	.	∼1	0.2	10^{34}	steady
15241 −617	1.4	dwarf?	≲1		∼7?	0.5?	10^{38}	transient
16088 −522	3–7	dwarf?	∼1		∼7?	0.1?	10^{38}	transient
16170 −155	∼5	emission	∼1	0.78	0.3–1	0.1–0.4	2×10^{37}	steady
16560 +354	≳15	A–F	∼2	1.70	∼3	1.5	10^{37}	steady
17051 −250	∼3	dwarf?	≲1		∼3	0.5	10^{38}	transient
17353 −444	5.2	emission	1–2		5–10	0.6–1.2	4×10^{37}	bursts, steady
18131 −140	4–5	G V	∼1		1.4	0.03	5×10^{36}	steady
19087 +005	5.5	K0 V	0.8	1.3	1.7	0.12	10^{38}	transient
19570 +115	?	F5–G	∼1		∼7	1.1	7×10^{36}	steady
20305 +407	∼3	dwarf?	≲1	0.2	≳10	0.12	≳10^{38}	steady
21426 +380	3.5	G?	∼1	9.8?	∼1	0.2	2×10^{36}	steady

Note: L_x for transient and burst sources corresponds to their maximum.

TABLE IV

X-ray pulsars

Source 2 XRS	Period (sec)	$P_{orb.}$ (days)	Distance (kpc)	Luminosity (erg s^{-1})
01152+634	3.615	24.30	5–7	transient, 3×10^{37}
01157−737	0.715	3.89	65	5×10^{38}
03522+308	835.80	580?	0.3	5×10^{33}
05357+262	103.83	−	2.5?	transient, 2×10^{37}
09002−403	282.90	8.96	1.3	7×10^{35}
11189−615	405.30	−	−	transient
11190−603	4.837	2.09	8	6×10^{37}
12238−624	695.7	35	2	4×10^{36}
12391−599	191?	−	−	−
12582−613	272.2	−	2	4×10^{35}
15382−521	528.93	3.79	6.5	2×10^{36}
16272−673	7.68	0.1?	−	−
16531−407	38.22	7.8?	1.4?	10^{35}?
16560+354	1.24	1.70	5	5×10^{36}
17005−377	5820	3.41	1.3	6×10^{35}
17289−247	122.61	−	−	−
18131−140	1912.8	−	−	−

TABLE V

Extragalactic sources

00058+200	Seyfert galaxy MKN 235	03165+413	cluster of galaxies Abell 426
00079+106	quasar III Zw 2	03210−450	cluster of galaxies
00092−339	cluster of galaxies SG 0000−303	03318−363	emission galaxy NGC 1365
00218+420	galaxy M31	03330+317	quasar NRAO 140
00264−730	extragalactic globular cluster Kron 3	03432−536	cluster of galaxies
00332+588	emission galaxy 3C 14.1	03492−139	quasar PKS 0349−1403
00390+411	see XRS 00218+420	04106+103	cluster of galaxies Abell 478
00397−096	cluster of galaxies Abell 85	04150+379	N-galaxy
00503−727	SMC X-3 (in small Magellanic cloud)	04234−531	cluster of galaxies NGC 1566
00529−739	SMC X-2 (in small Magellanic cloud)	04276−077	cluster of galaxies Abell 494
00549−015	cluster of galaxies Abell 119	04305+052	Seyfert galaxy 3C 120
01020−242	cluster of galaxies Abell 140, 141	04309−615	cluster of galaxies
01039−218	cluster of galaxies Abell 133	04312−136	cluster of galaxies Abell 496
01114−149	cluster of galaxies Abell 151	04461+449	cluster of galaxies 3C 129.1
01157−737	SMC X-1 (in small Magellanic cloud)	04574−357	group of galaxies
01206−591	Seyfert galaxy F9	05050−213	cluster of galaxies Abell 514
01209−353	emission galaxy NGC 526a	05174−456	N-galaxy Pic A
01430−330	cluster of galaxies SC 0141−340	05175+175	peculiar galaxy 3C 138
01486+360	cluster of galaxies Abell 262	05198+065	cluster of galaxies Abell 539
02065−019	Seyfert galaxy MKN 590	05212−365	object of type BL Lac
02285−130	cluster of galaxies Abell 358	05213−719	LMC X-2 (in Large Magellanic cloud)
02410+622	quasar and seyfert galaxy	05263−683	LMC X-5 (in Large Magellanic cloud)
02522−060	cluster of galaxies Abell 400	05270−350	cluster of galaxies
02530+417	cluster of galaxies Abell 396/397	05303−370	cluster of galaxies
02538+193	Seyfert galaxy MKN 372	05328−663	LMC X-4 (in Large Magellanic cloud)
02554+132	cluster of galaxies Abell 399/401	05357−668	burster in Large Magellanic cloud.
03162−443	cluster of galaxies PKS 0316−44	05370−441	Lacertid PKS 0537−441

Table V (continued)

05382 − 661	burster in LMC	14044 + 145	cluster of galaxies Abell 1852
05389 − 641	LMC X-3 (in Large Magellanic cloud)	14106 − 029	emission galaxy NGC 5506
05390 − 669	burster in Large Magellanic cloud	14156 + 255	Seyfert galaxy NGC 5548
05401 − 697	LMC X-1 (in Large Magellanic cloud)	14162 − 589	cluster of galaxies Abell 1415
05438 − 316	cluster of galaxies	14186 + 485	cluster of galaxies Abell 1904
05441 − 665	source in Large Magellanic cloud	14550 + 191	cluster of galaxies Abell 1991
05488 − 322	Lacertid	15087 + 062	cluster of galaxies Abell 2029
05497 − 074	emission galaxy NGC 2110	15142 + 068	cluster of galaxies Abell 2052
05512 + 466	Seyfert galaxy MCG 8–11–11	15190 + 082	galaxy NGC 5920
06261 − 541	cluster of galaxies	15212 + 285	cluster of galaxies Abell 2065
06384 + 742	Seyfert galaxy MKN 6.	15565 + 272	cluster of galaxies Abell 2142
06430 + 534	Seyfert galaxy anon. 0636 + 53	16011 + 159	cluster of galaxies Abell 2147, 2152
07080 + 357	cluster of galaxies	16212 − 234	cluster of galaxies Abell
07107 + 456	Seyfert galaxy MKN 376	16240 + 659	Seyfert galaxy MKN 876
07202 + 558	cluster of galaxies Abell 576	16278 + 396	cluster of galaxies Abell 2199
07380 + 498	Seyfert galaxy MKN 79	16284 + 286	cluster of galaxies Abell 2200
08152 − 075	cluster of galaxies Abell 644	16364 + 052	cluster of galaxies Abell 2204
09062 − 095	cluster of galaxies Abell 754	16516 + 399	Lacertid MKN 501
09433 − 140	interacting system NGC 2992/3	16592 + 337	cluster of galaxies Abell 2244
09459 − 306	cluster of galaxies JC 0948 − 327	17074 + 786	cluster of galaxies Abell 2256
09544 + 700	active galaxy M 82	17194 + 661	cluster of galaxies Abell 2255
09587 − 359	cluster of galaxies	17202 + 346	cluster of galaxies Abell 2261, 2262
10088 + 138	cluster of galaxies Abell 999	17271 + 503	Lacertid IZw 1727 + 502
10151 − 254	cluster of galaxies Abell 955	18010 + 698	N-galaxy 3C 371
10284 + 512	Seyfert galaxy MKN 142	18248 + 644	3C 383
10335 − 270	cluster of galaxies Abell 1060	18332 + 326	N-galaxy 3C 382
10586 − 226	cluster of galaxies Abell 1146	18341 − 626	Seyfert galaxy ESO 140–443
11020 + 384	emission galaxy MKN 421	18347 − 653	Seyfert galaxy ESO 103–G35
11304 − 146	cluster of galaxies Abell 1285	18476 + 789	N-galaxy 3C 390.3
11357 − 373	Seyfert galaxy NGC 3783	18543 + 683	cluster of galaxies A 2312
11440 + 197	cluster of galaxies Abell 1367	19146 − 589	Seyfert galaxy ESO 141–G55
11474 − 124	cluster of galaxies Abell 1391	19160 − 793	cluster of galaxies SC 1834 − 792
11500 + 720	cluster of galaxies Abell 1254	19197 + 436	cluster of galaxies Abell 2319
12078 + 397	Seyfert galaxy NGC 4151	19386 − 105	Seyfert galaxy NGC 6814
12190 + 305	Lacertid	19556 − 689	cluster of galaxies CA 1955 − 692
12260 + 024	quasar 3C 273	19572 + 405	radiogalaxy Cyg A
12287 + 126	cluster of galaxies Virgo A	20091 − 569	cluster of galaxies
12403 − 056	cluster of galaxies Abell 1588	20407 − 115	Seyfert galaxy MKN 509
12462 − 410	cluster of galaxies NGC 4696	21409 − 602	cluster of galaxies SC 2146 − 594
12544 − 169	cluster of galaxies Abell 1644	21559 − 304	Lacertid
12566 − 171	see XRS 12544 − 169	22095 − 471	Seyfert galaxy NGC 7213
12574 + 283	cluster of galaxies Abell 1656	22204 − 022	cluster of galaxies Abell 2440
13108 + 371	emission galaxy NGC 5005, 5033	22264 + 014	cluster of galaxies Abell 2457
13223 − 427	peculiar galaxy Cen A	22514 − 178	quasar MR 2251 − 178
13255 − 020	cluster of galaxies Abell 1750	22595 + 085	Seyfert galaxy NGC 7469
13264 + 119	cluster of galaxies Abell 1735	21320 − 214	cluster of galaxies Abell 2554
13268 − 311	cluster of galaxies SC 1329 − 314	23153 − 428	emission galaxy NGC 7582
13358 + 402	cluster of galaxies Abell 1763	23221 + 166	cluster of galaxies Abell 2589
13448 − 325	cluster of galaxies SC 1345 − 301	23440 + 086	cluster of galaxies Abell 2657
13464 − 300	quasar IC 4329 A	23448 − 285	cluster of galaxies K 44
13468 + 266	cluster of galaxies Abell 1795	25454 + 273	cluster of galaxies Abell 2666
13481 + 700	Seyfert galaxy MKN 279	23512 + 067	Seyfert galaxy MKN 541.
13505 + 390	Seyfert galaxy MKN 464		

TABLE VI
Transient

00421 + 327	16301 − 472
01152 + 634	17051 − 250
01322 + 007	17196 − 347
02140 + 045	17250 − 302
03362 + 010	17309 − 220
15357 + 262	17354 − 284
06201 − 003	17424 − 289
06560 − 071	17430 − 295
08362 − 426	17437 − 316
11189 − 615	17448 − 361
12466 − 588	17460 − 204
13240 − 625	18037 − 246
14104 − 619	18079 − 108
14484 − 556	18351 − 078
14553 − 313	19017 + 031
15241 − 617	19087 + 005
15438 − 475	19117 − 108
15539 − 542	19185 + 146
16088 − 522	21091 + 385
16243 − 490	21370 + 567

TABLE VII
Bursters (SB)

03110 + 420	17068 − 434
05124 − 400	17100 − 510
06143 + 051	17270 − 335
07360 − 500	17286 − 337
09020 + 573	17300 − 238
11508 + 745	17353 − 444
12170 − 672	17300 − 333
12270 + 190	17360 − 163
15168 − 569	17400 − 463
15170 − 500	17417 − 296
15190 − 314	17426 − 292
16040 − 590	17468 − 370
16088 − 522	18063 − 274
16190 − 468	18070 − 365
16243 − 490	18204 − 303
16369 − 536	18217 − 054
16420 − 343	18360 − 227
16450 − 507	18374 + 049
16450 − 575	18482 − 079
16460 + 284	19059 + 000
16589 − 289	19161 − 053
17051 − 431	19290 + 079
	19490 + 440

TABLE VIII
Bursters (LB)

00000 + 278
03531 − 400
04490 − 550
05000 − 554
06150 − 093
11016 + 384
12446 − 603
15358 − 292
17104 − 303
17161 − 318
17228 − 305
17428 − 294
17436 − 285
17441 − 284
18503 − 087
19082 + 023
23587 + 210
23588 − 640

TABLE IX
SN-remnants

00224 + 638	SN Tycho
02140 + 623	HB 3
04150 − 120	
05315 + 219	Crab
06137 + 227	IC 443
08215 − 427	Pup A
08336 − 450	Vela X
10450 − 593	G287.8–0.5
11222 − 590	MSH 11–54
12070 − 521	PKS 1209 − 52
14392 − 622	MSH 14–63
14580 − 415	SN 1006
15100 − 390	Lup Loop
15101 − 590	NSH 15–52
15400 − 320	
15452 − 536	NSH 15–56
16139 − 509	RCW 103
16186 + 150	North Polar Spur
17276 − 214	SN Kepler
18536 + 012	G34.6–0.5
19365 + 326	G63.6 + 6.0
20200 + 405	W 66
20497 + 308	Cyg Loop
23212 + 585	Cas A

TABLE X
Soft sources

01364 − 182	UV Cet	12200 + 260	
03064 + 477	LX Per	13083 + 362	HD 114519
03129 + 345		13140 + 292	white dwarf
03140 + 380			HZ 43
03210 + 236		13200 − 105	
03228 + 285	UX Ari	13522 + 187	η Boo
03342 + 002	HR 1099	14010 − 452	
05130 + 459	Capella	14260 − 624	Proxima Cen
05382 − 661	Bar in SMC	14362 − 606	α Cen
06225 − 529		14490 + 193	ξ Boo
06429 − 166	Sirius	16128 + 339	σ CrB
07394 + 036	UZ CMi	16220 − 243	
07401 + 290	HD 62044	16400 + 400	
07521 + 221	U Gem	17300 − 370	
08081 − 352	Nova Pup	17401 − 391	λ Sco
10270 + 590		18353 + 387	Vega
10520 + 560		18389 + 378	AU Lyr
10528 + 384	extragalactic	19010 + 430	
11015 + 450	AN UMa	20380 + 290	
11110 − 603		21200 + 450	
11353 + 525	RW UMa	22046 + 452	AR Lac
12150 + 440		22049 + 472	HK Lac
		23108 + 024	

TABLE XI
Hard

03250 + 440	·burst ∼ 12 min.
04360 + 120	burst ∼ 11 s.
05263 − 660	
11508 + 745	
17106 − 474	
17329 − 449	burst
18025 − 448	bursts
18061 + 458	

References

Abramenko, A., Gershberg, R., Pavlenko, B., Prokof'eva, V., Lewin, W., Hoffman, J., and Li, F.: 1978, preprint.
Agrawal, P. and Riegler, G.: 1978, preprint.
Agrawal, P. and Riegler, G.: 1978, *IAU Circ.* No. 3343.
Aksenov, E.: 1976, *IAU Circ.* No. 2946.
Allen, D.: 1978, *IAU Circ.* No. 3143.
Amnuel, P. R. and Guseinov, O. H.: 1979, *Astrophys. Space Sci.* **63**, 131.
Amnuel, P. R. and Guseinov, O. H.: 1980, *Astrophys. Space Sci.* **68**, 315.
Amnuel, P. R., Guseinov, O. H., and Rakhamimov, Sh. Yu.: 1974, *Astrophys. Space Sci.* **29**, 331.
Amnuel, P. R., Guseinov, O. H., and Rakhamimov, Sh. Yu.: 1979, *Astrophys. J. Suppl.* **41**, 327.
Andrew, B., Purton, C., Rappaport, S., Bradt, H., and Schnopper, H.: 1970, *Astrophys. J.* **161**, L173.
Apparao, K.: 1977, *Monthly Notices Roy. Astron. Soc.* **179**, 763.
Apparao, K., Bradt, H., Dower, R., Doxey, R., Jernigan, J., and Li, F.: 1978, *Nature* **271**, 225.
Apparao, K., Maraschi, L., Bigmaini, G., Helmken, H., Margon, B., Hjellming, R., Bradt, H., and Dower, R.: 1978a, *Nature* **273**, 450.

Arens, J. and Rothschild, R.: 1975, preprint.

Ashok, N., Kulkarni, P., Apparao, K., and Chitre, S.: 1978, *IAU Circ.* No. 3344.

Auriemma, G., Cardini, D., Costa, E., Ciovannelli, F., Orciuolo, H., and Ranieri, M.: 1976, *Nature* **259**, 27.

Auriemma, G., Angeloni, L., Belli, B., Bernardt, A., Cardini, D., Costa, E., Emanuele, A., Giovanelli, F., and Ubertini, P.: 1978, *Astrophys. J.* **221**, 7.

Avni, Y. and Bahcall, J.: 1975, *Astrophys. J.* **202**, L131.

Avni, Y. and Bahcall, J.: 1976, in *X-ray Binaries*, p. 615.

Bahcall, N.: 1976, *Astrophys. J.* **204**, L83.

Bahcall, N. and Hausman, M.: 1976, *Astrophys. J.* **207**, L181.

Bahcall, J., Charles, P., Davison, P., Sanford, P., Kellog, E., and York, D.: 1975, *Monthly Notices Roy. Astron. Soc.* **171**, 41 p.

Bahcall, N., Harries, D., and Strom, R.: 1976, *Astrophys. J.* **209**, L17.

Bahcall, N., Lasker, B., and Wamsteker, W.: 1977, *Astrophys. J.* **213**, L105.

Barat, S., Chambon, G., Hurley, K., Niel, M., Vedrenne, G., Estulin, I. V., Kurt, V. G., and Zenchenko, V. A.: 1979, *Astron. Astrophys.* **79**, L24.

Baratta, C., Viotti, R., and Altamore, A.: 1978, *Astron. Astrophys.* **65**, L21.

Barnden, L. and Francey, R.: 1969, *Proc. Astron. Soc. Austral.* **1**, 263.

Bartolini, C., Garnieri, A., Piccioni, A., Gianrande, A., and Giovanelli, F.: 1978, *IAU Circ.* No. 3167.

Basko, M. M., Goranski, V. P., Lutiy, V. M., Ruzan, L. L., Sunyayev, R. A., and Schugarov, S. Yu.: 1976, *Astron. Zhurnal (Pis'ma)* **2**, 539.

Becker, R., Pravdo, S., Serlemitsos, P., and Swank, J.: 1976a, *IAU Circ.* No. 2953.

Becker, R., Boldt, E., Holt, S., Pravdo, S., Rothschild, R., Serlemitsos, P., and Swank, J.: 1976b, *Astrophys. J.* **207**, L167.

Becker, R., Boldt, E., Holt, S., Pravdo, S., Rothschild, R., Serlemitsos, P., and Swank, J.: 1976c, *Astrophys. J.* **209**, L65.

Becker, R., Pravdo, S., Saba, J., and Serlemitsos, P.: 1977a, *IAU Circ.* No. 3039.

Becker, R., Boldt, E., Holt, S., Pravdo., S., Rothschild, R., Serlemitsos, P., Smith, B., and Swank, J.: 1977b, *Astrophys. J.* **214**, 879.

Becker, R., Swank, J., Boldt, E., Holt, S., Pravdo, S., Saba, J. and Serlemitsos, P.: 1977c, *Astrophys. J.* **216**, L11.

Becker, R., Robinson-Saba, J., Boldt, E., Holt, S., Pravdo, S., Serlemitsos, P., and Swank, J.: 1978, *Astrophys. J.* **224**, L113.

Becker, R., Rothschild, R., Boldt, E., Holt, S., Pravdo, S., Serlemitsos, P., and Swank, J.: 1978, *Astrophys. J.* **221**, 912.

Becker, R., Boldt, E., Holt, S., Pravdo, S., Robinson-Saba, J., Serlemitsos, P., and Swank, J.: 1979, *Astrophys. J.* **227**, L21.

Belian, R., Connor, J., and Evans, W.: 1976a, *IAU Circ.* No. 2569.

Belian, R., Connor, J., and Evans, W.: 1976b, *Astrophys. J.* **206**, L135.

Belian, R., Connor, J., and Evans, W.: 1976c, *Astrophys. J.* **207**, L33.

Berezhnoy, V., Grechko, G., Gubarev, A., Kirillov, E., Klimuk, P., Kurt, V., Sevastyanov, V., Titarchuk, L., Moskalenko, E., and Scheffer, E.: 1977, in *Cospar Space Research*, Vol. XVII.

Bernacca, P.: 1976, in *X-ray Binaries*, p. 101.

Bernacca, P., Bianchini, A., Walker, A., Backman, D., Canizares, C., van Paradijs, J., Hoffman, J., Doty, J., Marshall, H., Wheaton, W., Jernigan, J., and Lewin, W.: 1979, *Monthly Notices Roy. Astron. Soc.* **186**, 287.

Berthelsdorf, R. and Culhane, J.: 1979, *Monthly Notices Roy. Astron. Soc.* **187**, 17P.

Bieging, J. and Downes, D.: 1975, *Nature* **258**, 307.

Bidelman, W. and Sanduleak, N.: 1977, *IAU Circ.* No. 3146.

Blanco, V.: 1977, *IAU Circ.* No. 3039.

Bleach, R., Henry, R., Meekins, J., Fritz, G., Shulman, S., and Friedman, H.: 1975, *Astrophys. J.* **197**, L13.

Bleeker, J., Deerenberg, A., Heise, J., Yamashita, K., and Tanaka, Y.: 1973, *Nature Phys. Sci.* **241**, 55.

Boldt, E., Holt, S., Rothschild, R., and Serlemitsos, P.: 1976a, in *X-ray Binaries*, p. 113.

Boldt, E., Serlemitsos, P., Kaluzienski, L., Pravdo, S., Peacock, A., Elvis, M., Watson, M., and Pounds, K.: 1976b, in *X-ray Binaries*, p. 245.

Boley, F., Wolfson, R., Bradt, R., Doxsey, R., Jernigan, J., and Hiltner, W.: 1976, *Astrophys. J.* **203**, L13.

Bolton, C.: 1976, in *X-ray Binaries*, p. 465.

Bonnet-Bidaud, J. and van der Keis, M.: 1979, *Astron. Astrophys.* **73**, 90.

Bopp, B. and Talcott, J.: 1978, *Astron. J.* **83**, 1517.

Bord, D., Mook, D., Petro, L., and Hiltner, W.: 1976, in *X-ray Binaries*, p. 677.

Bortle, J.: 1976, *IAU Circ.* No. 2935, 2942.

Bowyer, S., Byram, E., Chubb, T., and Friedman, H.: 1975, *Science* **147**, 394.

Bradt, H.: 1978, preprint.

Bradt, H.: 1979, preprint.

Bradt, H. and Matilsky, T.: 1976a, in *X-ray Binaries*, p. 317.

Bradt, H., Burnett, B., Meyer, W., Rappaport, S., and Schnopper, H.: 1971, *Nature* **229**, 96.

Bradt, H., Meyer, W., Buff, J., Clark, G., Doxsey, R., Hearn, D., Jernigan, J., Joss, P., Laufer, B., Lewin, W., Li, F., Matilsky, T., McClintock, J., Primini, F., Rappaport, S., and Schnopper, H.: 1976b, *Astrophys. J.* **204**, L67.

Bradt, H., Apparao, K., Clark, G., Dower, R., Doxsey, R., Hearn, D., Jernigan, J., Joss, P., Meyer, W., McClintock, J., and Walter, P.: 1977a, preprint.

Bradt, H., Apparao, K., Dower, R., Doxsey, R., Jernigan, J., and Markert, T.: 1977b, preprint.

Bradt, H., Doxsey, R., and Jernigan, J.: 1978, preprint.

Bradt, H., Doxsey, R., Johnston, M., Schwartz, D., Burkhead, M., Dent, W., Liller, W., and Smith, A.: 1979, *Astrophys. J.* **230**, L5.

Bradt, H., Burke, B., Canizares, C., Greenfield, P., Kelley, R., McClintock, J., van Paradijs, J., and Koski, A.: 1978, *Astrophys. J.* **226**, L111.

Braes, L., Miley, G., and Schoenmaker, A.: 1972, *Nature* **236**, 392.

Branduardi, G., Ives, J., Sanford, P., Brinkman, A., and Maraschi, L.: 1976, *Monthly Notices Roy. Astron. Soc.* **175**, 47.

Branduardi, G., Mason, K., Sanford, P.: 1978, *Monthly Notices Roy. Astron. Soc.* **185**, 137.

Bratoliubova-Tsulukidze, L. B., Kudriavtsev, M. M., Melioranski, A. S., Savekko, I. A., and Yuschkov, B. Yu.: 1976a, *Astron. Circ.* (Russian) No. 898.

Bratoliubova-Tsulukidze, L. B., Kudriavtsev, M. M., Melioranski, A. S., Savenko, I. A., and Yuschkov, B. Yu.: 1976b, *Kosmich. Issledov.* **14**, 641.

Brecher, K. and Wasserman, I.: 1974, *Astrophys. J.* **192**, L125.

Brini, D., Kiriegi, U., Fuligni, R., Moretti, E., and Vespignani, J.: 1967, *Astrophys. J.* **149**, 420.

Brini, D., Frontera, P., Fuligni, R., and Moretti, E.: 1976, *Astrophys. J.* **18**, L15.

Bunner, A.: 1978a, *IAU Circ.* No. 3339.

Bunner, A. and Palmieri, T.: 1969, *Astrophys. J.* **158**, L35.

Bunner, A. and Sanders, W.: 1976, in *X-ray Binaries*, p. 187.

Bunner, A. and Sanders, W.: 1979, *Astrophys. J.* **228**, L19.

Burke, J.: 1976, *Astrophys. J.* **209**, 556.

Byrne, P. and Wayman, P.: 1977, *Monthly Notices Roy. Astron. Soc.* **178**, 45.

Cahn, J.: 1977, *Astron. Astrophys.* **58**, 443.

Calla, O., Bhandari, S., Deshpande, M., Vatshari, O.: 1978, *IAU Circ.* No. 3347.

Campisi, I., Traves, A., and Bernacca, P.: 1976, *Monthly Notices Roy. Astron. Soc.* **176**, 225.

Canizares, C. and Oda, M.: 1977, *Astrophys. J.* **214**, L119.

Canizares, C., Bradt, H., Buff, J., and Laufer, B.: 1976, in *X-ray Binaries*, p. 373.

Canizares, C., McClintock, J., and Ricker, G.: 1978a, preprint.

Canizares, C., Grindlay, J., Hiltner, W. A., Miller, W., and McClintock, J.: 1978, *Astrophys. J.* **224**, 39.

Canizares, C., McClintock, J., and Grindlay, J.: 1979, preprint.

Carpenter, G., Eyles, C., Skinner, G., Willmore, A., and Wilson, A.: 1975, *IAU Circ.* No. 2852.

Carpenter, G., Eyles, C., Skinner, G., Wilson, A., and Willmore, A.: 1977, *Monthly Notices Roy. Astron. Soc.* **179**, 27P.

Cash, W., Charles, P., Bowyer, S.: 1979, *Astron. Astrophys.* **72**, L6.

Catura, R., Acton, L., and Johnson, H.: 1975, *Astrophys. J.* **196**, L47.

Catura, R. and Acton, L.: 1976a, *Astrophys. J.* **207**, L163.

Catura, R. and Acton, L., 1976b, in *X-ray Binaries*, p. 119.

Charles, P., Culhane, J., and Zarnecki, J.: 1975a, *Astrophys. J.* **196**, L19.

Charles, P., Thorstensen, J., Bowyer, S., Middleditch, J.: 1979, *Astrophys. J.* **231**, L131.

Charles, P., Culhane, J., Fabian, A., Mitchell, R., and Zarnecki, J., 1975b, *Monthly Notices Roy. Astron. Soc.* **170**, 61P.

Charles, P., Mason, K., Culhane, J., Sanford, P., and White, N.: 1976, in *X-ray Binaries*, p. 629.

Charles, P., Mason, K., White, N., Culhane, J., Sanford, P., and Moffat, A.: 1978, *Monthly Notices Roy. Astron. Soc.* **183**, 813.

Charles, P., Thorstensen, J., and Bowyer, S.: 1978b, *Monthly Notices Roy. Astron. Soc.* **183**, 29.

Charles, P., Holt, S., and Sanford, P.: 1978c, *IAU Circ.* No. 3235.

Chevalier, C., Ilovaisky, S., Branduardi, G., and Sanford, P.: 1976a, in *X-ray Binaries*, p. 717.

Chevalier, C., Ilovaisky, S., and Mauder, H.: 1976b, *IAU Circ.* No. 2957.

Chevalier, C. and Ilovaisky, S.: 1977, *IAU Circ.* No. 3073.

Chi-Chao, W., van Duinen, R., and Hammerschlag-Hensberge, G.: 1976a, in *X-ray Binaries*, p. 529.

Chi-Chao, W., Anders, J., van Duinen, R., Kester, D., and Wesselius, P.: 1976b, *Astron. Astrophys.* **50,** 445.

Chi-Chao, W.: 1978, *Astrophys. J.* **227,** 291.

Chodil, J., Mark, H., Rodhues, R., Seward, F., Swift, C., and Hiltner, H.: 1967, *Phys. Rev. Letters* **19,** 681.

Ciatti, F., Mammano, A., and Vittone, A., 1977, *Astron. Astrophys.* **56,** 311.

Citterio, O., Conti, G., di Benedetto, P., Tanzi, E., Parola, G., White, H., Charles, P., and Sanford, P.: 1976, *Monthly Notices Roy. Astron. Soc.* **175,** 35.

Clark, G.: 1975a, *IAU Circ.* No. 2799.

Clark, G.: 1975b, *IAU Circ.* No. 2843.

Clark, G.: 1976, *IAU Circ.* No. 2922.

Clark, G., Markert, T., and Li, F.: 1975a, preprint.

Clark, G., Jernigan, J., Bradt, H., Canizares, C., Lewin, W., Li, F., Mayer, W., and McClintock, J.: 1976, *Astrophys. J.* **207,** L105.

Clark, G. and Li, F.: 1977, preprint.

Clark, G. and Lewin, W.: 1977, 'Circ. Lett. Coord. Campaign to Obs'., *X-ray Binaries* No. 26.

Clark, G. and Chartres, M.: 1976, *IAU Circ.* No. 3208.

Clark, G., Doxsey, R., Li, F., and Marshall, F.: 1977, preprint.

Clrak, G., Doxsey, R., Li, F., Jernigan, J., and van Paradijs, J.: 1978, *Astrophys. J.* **221,** L37.

Clark, D., Parkinson, J., and Caswell, J.: 1975b, *Nature* **254,** 674.

Cline, T.: 1979, preprint TM-80630.

Cline, T., Desai, U., Pizzichini, G., Teegarden, B., Evans, W., Klebesadel, R., Laros, J., Hurlly, K., Niel, M., Vedrenne, G., Estoolin, I., Kouznetsov, A., Zenchenko, V., Hovestadt, D., and Gloeckler, G.: 1979, preprint IM-80570.

Coe, M., Engel, A., and Quenby, J.: 1976a, *Monthly Notices Roy. Astron. Soc.* **177,** 31P.

Coe, M., Engel, A., and Quenby, J.: 1976b, *Nature* **262,** 563.

Coe, M., Engel, A., and Quenby, J.: 1976c, *IAU Circ.* No. 3003.

Coe, M., Engel, A., and Quenby, J.: 1977, *IAU Circ.* No. 3054.

Coe, M., Carpenter, G., Engel, A., and Quenby, J.: 1975, *Nature* **256,** 630.

Coleman, P., Bunner, A., Kraushaar, W., and McCammon, D.: 1971, *Astrophys. J.* **170,** L47.

Cominsky, L., Forman, W., Jones, C., and Tananbaum, H.: 1977, *Astrophys. J.* **211,** L9.

Cominsky, L., Jones, C., Forman, W., and Tananbaum, H.: 1978, *Astrophys. J.* **221,** 46.

Cominsky, L., Clark, G., Li, F., Mayer, W., and Rappaport, S.: 1978a, preprint.

Cominsky, L., Li, F., Bradt, H., and Rappaport, S.: 1978b, *IAU Circ.* No. 3163.

Conner, J., Evans, W., and Belian, R.: 1969, *Astrophys. J.* **157,** L157.

Cooke, B.: 1976, *Nature* **261,** 564.

Cooke, B. and Pounds, K.: 1971, *Nature Phys. Sci.* **229,** 144.

Cooke, B. and Ricketts, M.: 1976, *IAU Circ.* No. 2977.

Cooke, B. and Maccagni, B.: 1976, *Monthly Notices Roy. Astron. Soc.* **175,** 65P.

Cooke, B., Lawrence, A., and Perola, G.: 1978a, *Monthly Notices Roy. Astron. Soc.* **182,** 661.

Cooke, B., Ricketts, M., Maccacaro, T., Pye, J., Elvis, M., Watson, M., Griffiths, R., Pounds, K., McHardy, I., Maccagni, D., Seward, F., Page, C., and Turner, M.: 1978b, *Monthly Notices Roy. Astron. Soc.* **182,** 489.

Cordova, F. and Riegler, G.: 1978, preprint.

Cordova, F. and Garmire, G.: 1978, preprint.

Cordova, F., Garmire, G., and Tuohy, I.: 1978, *IAU Circ.* No. 3235.

Cowley, A. and Crampton, D.: 1975, *Astrophys. J.* **201,** L65.

Cowley, A., Crampton, D., Szkody, P., and Brownlee, D.: 1976a, *IAU Circ.* No. 2984.

Cowley, A., Rogers, L., and Hutchings, J.: 1976b, *Publ. Astron. Soc. Pacific* **88,** 911.

Crampton, D.: 1978, *IAU Circ.* No. 3317.

Crampton, D. and Cowley, A.: 1976, *Astrophys. J.* **207,** L171.

Crampton, D. and Cowley, A.: 1978, *IAU Circ.* No. 3292.

Crampton, D., Cowley, A., Hutchings, J., and Kaat, C.: 1976, *Astrophys. J.* **207,** 907.

Crampton, D., Hutchings, J., and Cowley, A.: 1978, *Astrophys. J.* **225,** L63.

Cruddace, R.: 1971, preprint.

Cruddace, R., Bowyer, S., Lampton, M., Mack, J., and Margon, B.: 1972, *Astrophys. J.* **174,** 529.

Cruddace, R., Friedman, H., Fritz, C., and Shulman, S.: 1976, *Astrophys. J.* **207**, 888.

Culhane, J.: 1977, in D. Schramm (ed.), *Supernovae*, p. 29.

Culhane, J., Mason, K., Sanford, P., and White, N.: 1976, in *X-ray Binaries*, p. 1.

Danks, A., Wamsteker, W., Vogt, N., Salinori, P., Tarenghi, M., and Duerbeck, H.: 1979, *Astrophys. J.* **227**, L59.

Davidsen, A., Malina, R., Smith, H., Spinrad, H., Margon, B., Mason, K., Hawkins, F., and Sanford, P.: 1974, *Astrophys. J.* **193**, L25.

Davidsen, A., Malina, R., and Bowyer, S.: 1976, *Astrophys. J.* **203**, 448.

Davidsen, A., Henry, R., Snyder, W., Friedman, R., Fritz, G., Naranan, S., Shulman, S., and Yentis, D.: 1977, *Astrophys. J.* **215**, 541.

Davies, R., Edwards, M., Morrison, I., and Spencer, R.: 1975, *Nature* **257**, 659.

Davies, R., Walsh, D., Browns, I., Edwards, M., and Noble, R.: 1976, *Nature* **261**, 476.

Davison, P.: 1977a, *Monthly Notices Roy. Astron. Soc.* **179**, 15P.

Davison, P.: 1977b, *Monthly Notices Roy. Astron Soc.* **179**, 35P.

Davison, P.: 1977c, *IAU Circ.* No. 3047.

Davison, P.: 1977d, *IAU Circ.* No. 3078.

Davison, P.: 1978, *Monthly Notices Roy. Astron. Soc.* **183**, 39.

Davison, P. and Tuohy, J.: 1975, *Monthly Notices Roy. Astron. Soc.* **173**, 33P.

Davison, P., Burnell, J., Ives, J., Wilson, A., and Carpenter, G.: 1976, *IAU Circ.* No. 2925.

Davison, P. and Morrison, L.: 1977, *Monthly Notices Roy. Astron. Soc.* **178**, 53P.

Davison, P., Watson, M., and Pye, J.: 1977, *Monthly Notices Roy. Astron. Soc.* **181**, 73P.

Dolan, J.: 1970, *Astron. J.* **75**, 223.

Dorren, J., Guinan, E., and McCook, G.: 1978, *IAU Circ.* No. 3352.

Doty, J.: 1976, *IAU Circ.* Nos. 2910, 2922.

Dower, R. and Kelley, R.: 1977, *IAU Circ.* No. 3144.

Doxsey, R., Jernigan, J., Hearn, D., Bradt, H., Buff, J., Clark, G., Delvaille, J., Epstein, A., Joss, P., Matilsky, T., Mayer, W., McClintock, J., Rappaport, S., Richardson, J., and Scnopper, H.: 1976, *Astrophys. J.* **203**, L9.

Doxsey, R., Apparao, K., Bradt, H., Dower, R., and Jernigan, J.: 1977, *Nature* **269**, 112.

Doxsey, R., Grindlay, J., Griffiths, R., Bradt, H., Johnston, M., Leach, R., Schwartz, D., and Schwartz, J.: 1978, *Astrophys. J.* **228**, L67.

Duerbeck, H. and Walter, K.: 1976, *Astron. Astrophys.* **48**, 141.

Dupree, A. and Lester, J.: 1976, in *X-ray Binaries*, p. 539.

Eachus, L., Wright, E., and Liller, W.: 1976, *Astrophys. J.* **203**, L17.

Elvis, M., Page, C., Pounds, K., Ricketts, M., and Turner, M.: 1975, *Nature* **257**, 656.

Epstein, A.: 1975, *IAU Circ.* No. 2859.

Epstein, A.: 1977, *Astrophys. J.* **218**, L19.

Eyles, C., Skinner, G., Willmore, A., and Rosenberg, F.: 1975a, *Nature* **254**, 577.

Eyles, C., Skinner, G., Willmore, A., and Rosenberg, F.: 1975b, *Nature* **254**, 291.

Evans, W., Belian, R., and Conner, J.: 1970, *Astrophys. J.* **159**, L57.

Evans, W., Belian, R., and Conner, J.: 1976, *Astrophys. J.* **207**, L91.

Fabbiano, G. and Schreier, E.: 1977, *Astrophys. J.* **214**, 235.

Fabbiano, G., Bradt, H., Doxsey, R., Gursky, H., Schwartz, D., and Schwarz, J.: 1978, *Astrophys. J.* **221**, L49.

Fabbiano, G., Gursky, H., Schwartz, D., Schwarz, J., Bradt, H., and Doxsey, R.: 1978a, preprint.

Fabbiano, G., Doxsey, R., Johnston, M., Schwartz, D., and Schwarz, J.: 1979, *Astrophys. J.* **230**, L67.

Fechner, W. and Joss, P.: 1977, *Astrophys. J.* **213**, L57.

Ferrary-Toniolo, M., Natali, G., Persi, P., and Spada, G.: 1977, *Astron. Astrophys.* **61**, 47.

Ferrari-Toniolo, M., Persi, P., and Viotti, V.: 1978, *Monthly Notices Roy. Astron. Soc.* **185**, 841.

Ficher, P., Jordan, W., Meyerott, A., Acton, L., and Roething, D.: 1966, *Nature* **211**, 920.

Fisher, P., Jordan, W., Meyerott, A., Acton, L., and Roething, D.: 1968, *Astrophys. J.* **151**, 1.

Forman, W. and Liller, W.: 1973, *Astrophys. J.* **183**, L117.

Forman, W. and Jones, C.: 1976, *Astrophys. J.* **207**, L177.

Forman, W., Jones, C., and Tananbaum, H.: 1976a, *Astrophys. J.* **206**, L29.

Forman, W., Jones, C., and Tananbaum, H.: 1976b, *Astrophys. J.* **207**, L25.

Forman, W., Jones, C., and Tananbaum, H.: 1976c, *Astrophys. J.* **208**, 849.

Forman, W., Jones, C., Cominsky, L., Julien, P., Murray, S., Peters, G., Tananbaum, H., and Giacconi, R.: 1978, *Astrophys. J. Suppl.* **38**, 357.

Forman, W., Jones, C., Murray, S., and Giacconi, R.: 1978a, *Astrophys. J.* **225**, L1.

French, H.: 1975, *IAU Circ.* No. 2835.

Friedman, H., Byram, E., and Chubb, T.: 1967, *Science* **156**, 374.

Fritz, G., Davidsen, A., Meekins, J., and Friedman, H.: 1971, *Astrophys. J.* **164**, L81.

Fritz, G., Naranan, S., Shulman, S., Yentis, D., Friedman, H., Davidsen, A., Henry, R., and Snyder, W.: 1976, *Astrophys. J.* **207**, L29.

Frontera, F. and Fuligni, F.: 1975, *Astrophys. J.* **196**, 597.

Frontera, F. and Fuligni, F.: 1976, *Astrophys. Space Sci.* **42**, 185.

Frontera, F., Fuligni, F., Morelli, E., and Ventura, G.: 1979, *Astrophys. J.* **229**, 291.

Fujimoto, M., Hayakawa, S., and Kato, T.: 1969, *Astrophys. Space Sci.* **4**, 64.

Fuligni, F., Brini, D., Dusi, W., and Frontera, F.: 1976, *Astrophys. J.* **208**, L111.

Galas, C., Tuohy, I., and Garmire, G.: 1978, *IAU Circ.* No. 3254.

Galper, A. M., Kirillov-Ugriumov, V. G., Kurochkina, A. V., Leikov, N., Luchkov, B. I., and Yurkin, Ya. T.: 1976, *Pis'ma v Astron. J.* (Russian), **2**, 524.

Garmire, G., Charles, P., Mason, K., Bowyer, S., Riegler, G., Tuohy, I., Boldt, E., Holt, S., Rothschild, R., and Serlemitsos, P.: 1977, *IAU Circ.* No. 3125.

Giacconi, R., Gursky, H., and Waters, J.: 1965, *Nature* **207**, 572.

Giacconi, R., Murray, S., Gursky, H., Kellogg, E., Schreier, E., Matilsky, T., Kock, D., and Tananbaum, H.: 1974, *Astrophys. J. Suppl.* **27**, 37.

Giangrande, A., Giovanelli, F., Bartolini, C., Guarnieri, A., and Piccioni, A.: 1978 *IAU Circ.* No. 3129.

Glass, I.: 1978, *Monthly Notices Roy. Astron. Soc.* **183**, 335.

Glass, I.: 1979, *Monthly Notices Roy. Astron. Soc.* **187**, 807.

Glass, J. and Feast, M.: 1973, *Nature Phys. Sci.* **245**, 39.

Glass, J.: 1976, *IAU Circ.* No. 2974.

Glass, I. and Feast, M.: 1978, *IAU Circ.* No. 3226.

Gorenstein, P.: 1969, preprint.

Gorenstein, P. and Giacconi, R.: 1968, preprint.

Gorenstein, P. and Tucker, W.: 1972, *Astrophys. J.* **176**, 333.

Gorenstein, P. and Tucker, W.: 1975, preprint.

Gorenstein, P., Harnden, F., and Tucker, W.: 1974, *Astrophys. J.* **192**, 661.

Goss, W., Haynes, R., Watkinson, A., and Skellern, D.: 1976, *IAU Circ.* No. 3003.

Gottlieb, E.: 1975, *Astrophys. J.* **202**, L13.

Gowen, R., Cooke, B., Griffiths, R., and Ricketts, M.: 1977, *Monthly Notices Roy. Astron. Soc.* **179**, 303.

Greenstein, J., Oke, J., and Wade, R.: 1978, *IAU Circ.* No. 3224.

Griffiths, R., Bradt, H., Doxsey, R., Friedman, H., Gursky, H., Johnston, M., Longmore, A., Malin, D., Murdin, D., Murdin, P., Schwartz, D., and Schwarz, J.: 1977, preprint.

Griffiths, R., Briel, U., Schwartz, D., Schwarz, J., Doxsey, R., and Johnstone, M.: 1979, preprint.

Griffiths, R., Bradt, H., Briel, U., Davyer, R., Doxsey, R., Fabbiano, G., Gursky, H., Johnston, M., Leach, R., Murdin, P., Ramsey, A., and Schwarz, J.: 1978, preprint.

Griffiths, R., Gursky, H., Schwartz, D., Schwarz, J., Bradt, H., Doxsey, R., Charles, P., and Thorstensen, J.: 1978a, *Nature* **276**, 247.

Griffiths, R., Doxsey, R., Johnston, M., Schwartz, D., Schwarz, J., and Blades, J.: 1979a, preprint.

Griffiths, R., Ward, M., Blades, J., and Wilson, A.: 1978b, *IAU Circ.* No. 3326.

Griffiths, R., Doxsey, R., Johnston, M., Schwartz, D., Schwarz, J., and Blades, J.: 1979b, *Astrophys. J.* **230**, L21.

Griffiths, R., Bradt, H., Doxsey, R., Friedman, H., Gursky, H., Johnston, M., Longmore, A., Malin, D., Murdin, P., Schwartz, D., and Schwarz, J.: 1978b, *Astrophys. J.* **221**, L63.

Griffiths, R., Tapia, S., Briel, U., and Chaisson, L.: 1979c, preprint.

Grindlay, J.: 1976, preprint.

Grindlay, J.: 1977, *IAU Circ.* No. 3104.

Grindlay, J.: 1978, *Astrophys. J.* **224**, L107.

Grindlay, J.: 1978a, preprint.

Grindlay, J.: 1978b, *Astrophys. J.* **225**, 1001.

Grindlay, J.: 1978c, *IAU Circ.* No. 3229.

Grindlay, J.: 1978d, *IAU Circ.* No. 3238.

Grindlay, J. and Gursky, H.: 1976a, *IAU Circ.* No. 2932.

Grindlay, J. and Gursky, H.: 1976b, *Astrophys. J.* **209**, L61.

Grindlay, J. and Liller, W.: 1978, *Astrophys. J.* **220**, L127.

Grindlay, J., Gursky, H., Parsignault, D., Cohn, H., Heise, J., and Brinkman, A.: 1977a, preprint.

Grindlay, J., Parsignault, D., Gursky, H., Brinkman, A., Heise, J., and Harries, D.: 1977b, *Astrophys. J.* **214**, L57.

Grindlay, J., McClintock, J., Canizares, C., van Paradijs, J., Li, F., and Lewin, W.: 1978, *Nature* **274,** 567

Grindlay, J., McClintock, J., Canizares, C., and van Paradijs, J.: 1978a, *IAU Circ.* No. 3230.

Gronenschild, E., Mewe, R., Heise, J., den Boggende, A., Schrijver, J., and Brinkman, A.: 1978, *Astron. Astrophys.* **65,** L9.

Groote, D., Kaufmann, J., and Hunger, K.: 1978, *Astron. Astrophys.* **63,** L9.

Guinan, E., McCook, G., and Derren, J.: 1978, *IAU Circ.* No. 3182.

Guinan, E., McCook, G., and Dorren, J.: 1977, *IAU Circ.* No. 3176.

Gull, T., York, D., Snow, T., and Henize, K.: 1976, *Astrophys. J.* **206,** 260.

Gursky, H. and Schreier, E.: 1975, in 'Variable Stars and Stellar Systems', *IAU Symp.* **67.**

Gursky, H., Gorenstein, P., and Giacconi, R.: 1967, *Astrophys. J.* **150,** L75.

Guseinov, O. H. and Kasumov, F. K.: 1978, *Astrophys. Space Sci.* **59,** 285.

Hackwell, J., Gehrz, R., Grasdalen, G., Cominsky, L., van Paradijs, J., and Lewin, W.: 1978, *IAU Circ.* No. 3331.

Hackwell, J., Grasdalen, G., Gehrz, R., van Paradijs, J., Cominsky, L., and Lewin, W.: 1979, preprint.

Haisch, B., Linsky, J., Slee, O., Hearn, D., Walker, A., Rydgren, A., and Nicolson, G.: 1978, *Astrophys. J.* **225,** L35.

Hammerschlag-Hensberge, G. and Wu, C.: 1977, *Astron. Astrophys.* **56,** 433.

Haynes, R., Jauncey, D., Murdin, P., Goss, W., Longmore, A., Simons, L., Milne, D., and Skeelern, D.: 1978, *Monthly Notices Roy. Astron. Soc.* **189,** 661.

Harries, J., Tuohy, J., Broderick, A., Fenton, K., and Luyendy, A.: 1971, *Nature Phys. Sci.* **234,** 149.

Hawkins, F. and Sanford, P.: 1976, *Mem. Sos. Astron. Ital.* **45,** 851.

Hayakawa, S., Kato, T., Nagase, F., Yamashita, I., Murakami, T., and Tanaka, Y.: 1977, *Astrophys. J. Letters* **213,** L109.

Haymes, R., Ellis, G., Fishman, G., Glenn, S., and Kurfess, J.: 1969, *Astrophys. J.* **157,** 1455.

Haymes, R., Caswell, J., and Simons, L.: 1977, *IAU Circ.* No. 2977.

Hearn, D.: 1976, *IAU Circ.* No. 2925.

Hearn, D.: 1978, *IAU Circ.* No. 3326.

Hearn, D. and Marshall, F.: 1979, *Astrophys. J.* **232,** L21.

Hearn, D., Richardson, G., and Clark, G.: 1976a, *Astrophys. J.* **210,** L23.

Hearn, D., Richardson, G., Bradt, H., Clark, G., Lewin, W., Mayer, W., McClintock, J., Primini, F., and Rappaport, S.: 1976b, *Astrophys. J.* **203,** L21.

Hearn, D., Marshall, F., and Jernigan, J.: 1979, *Astrophys. J.* **227,** L63.

Heise, J. and Grindlay, J.: 1976, *IAU Circ.* No. 2929.

Heise, J. and Brinkman, A.: 1976, in *X-ray Binaries*, p. 27.

Heise, J., Brinkman, A., Schrijver, J., Mewe, R., Gronenschild, E., den Boggende, A., and Grindlay, J.: 1975, *Astrophys. J.* **202,** L73.

Heise, J., Brinkman, A., den Boggende, A., Parsignault, D., Grindlay, J., and Gursky, H.: 1976, preprint.

Heise, J., Mewe, R., Brinkman, A., Gronenschild, E., den Boggende, A., Schrijver, J., Parsignault, D., and Grindlay, J.: 1978, *Astron. Astrophys.* **63,** L1.

Helmken, H., Delvaille, J., Epstein, A., Geller, M., and Schnopper, H.: 1978, *Astrophys. J.* **221,** 43.

Henry, P. and Tucker, W.: 1979, *Astrophys. J.* **229,** 78.

Henry, R. and Schreier, E.: 1977, *Astrophys. J.* **212,** L13.

Henry, R., Fritz, G., Meekins, J., Friedman, H., and Byram, E.: 1968, *Astrophys. J.* **153,** L11.

Hill, R., Byrginyon, G., Grader, R., Palmieri, T., Seward, F., and Stoering, J.: 1972, *Astrophys. J.* **171,** 519.

Hill, R., Burginyon, G., Grader, R., Toor, A., Stoering, J., and Seward, F.: 1974, *Astrophys. J.* **189,** L69.

Hiltner, W.: 1978, *IAU Circ.* No. 3324.

Hintzen, P. and Scott, J.: 1979, *Astron. Astrophys.* **74,** 116.

Hjellming, H.: 1976, in *X-ray Binaries*, p. 495.

Hoag, A. and Wiesberg, J.: 1976, *Astrophys. J.* **209,** 908.

Hoffman, J.: 1976, *IAU Circ.* No. 2946, 2953.

Hoffman, J., Lewin, W., Doty, J., Hearn, D., Clark, G., Jernigan, J., and Li, F.: 1976a, *Astrophys. J.* **210,** L13.

Hoffman, J., Doty, J., and Lewin, W.: 1976b, *IAU Circ.* No. 3025.

Hoffman, J., Lewin, W., and Doty, J.: 1977, *Monthly Notices Roy. Astron. Soc.* **179,** 57P.

Hoffman, J., Lewin, W., Primini, F., Wheaton, W., Swank, J., Boldt, E., Holt, S., Serlemitsos, P., Share, G., Wood, K., Yentis, D., Evans, W., Matteson, J., Gruber, D., and Peterson, L.: 1979, preprint.

Hoffman, J., Wheaton, W., Primini, F., Campbell, P., Dobson, C., Howe, S., Tziang, E., Scheepmaker, A., Lewin, W., Matteson, J., Gruber, D., Peterson, L., Swank, J., Boldt, E., Holt, S., Rothschild, R., and Serlemitsos, P.: 1978a, preprint.

Holt, S. and Kaluzienski, L.: 1977, *IAU Circ.* No. 3031.

Holt, S., Boldt, E., Serlemitsos, P., Murray, S., Giacconi, R., Kellogg, E., and Matilsky, T.: 1974, *Astrophys. J.* **188**, L97.

Holt, S., Boldt, E., Serlemitsos, P., and Kaluzienski, L.: 1976a, *Astrophys. J.* **205**, L27.

Holt, S., Kaluzienski, L., Boldt, E., and Serlemitsos, P.: 1976b, *Nature* **261**, 213.

Holt, S., Kaluzienski, L., Boldt, E., and Serlemitsos, P.: 1979, *Astrophys. J.* **227**, 503.

Holt, S., Kaluzienski, L., Mushotzky, R., Boldt, E., and Serlemitsos, P.: 1978, *IAU Circ.* No. 3292.

Holt, S., Kaluzienski, L., Boldt, E., and Serlemitsos, P.: 1979, *Astrophys. J.* **227**, 563.

Hutchings, J.: 1974, *Astrophys. J.* **192**, 677.

Hutchings, J.: 1975, *Astrophys. J.* **201**, 413.

Hutchings, J.: 1977, *Monthly Notices Roy. Astron. Soc.* **181**, 619.

Hutchings, J.: 1978, *Astrophys. J.* **226**, 264.

Hutchings, J., Cowley, A., Crampton, D., van Paradijs, J., and White, N.: 1979, *Astrophys. J.* **229**, 1079.

Ilovaisky, S., Chevalier, C., and Motch, C.: 1978, *Astron. Astrophys.* **71**, L17.

Ilovaisky, S., Motch, C., and Chevalier, C.: 1978a, *IAU Circ.* No. 3233.

Ilovaisky, S., Motch, C., and Chevalier, C.: 1978b, *Astron. Astrophys.* **70**, L19.

Ilovaisky, S., Chevalier, C., Motch, C., and Janot-Pacheco, E.: 1978c, *IAU Circ.* No. 3325.

Ives, J., Sanford, P., and Burnell, S.: 1975, *Nature* **254**, 578.

Jain, A., Jayanthi, U., Kasturirangan, K., and Rao, U.: 1976, *Astrophys. Space Sci.* **45**, 433.

Jensen, K.: 1979, *IAU Circ.* No. 3382.

Jernigan, J.: 1976, *IAU Circ.* No. 2900.

Jernigan, J.: 1976a, *IAU Circ.* No. 2957.

Jernigan, J. and Clark, G.: 1979, *Astrophys. J.* **231**, L125.

Jernigan, J., Apparao, K., Bradt, H., Doxsey, R., Dower, R., and McClintock, J.: 1978, preprint.

Jernigan, J., Bradt, H., van Paradijs, J., and Rappaport, S.: 1978, *IAU Circ.* No. 3225.

Jernigan, J., Apparao, K., Bradt, H., Doxsey, R., Dower, R., and McClintock, J.: 1978a, preprint.

John, R.: 1976, *Astrophys. J.* **208**, L31.

Johnson, H.: 1978, *IAU Circ.* No. 3184.

Johns, M., Koski, A., Canizares, C., and McClintock, J.: 1978, *IAU Circ.* No. 3171.

Johnston, M., Bradt, H., Doxsey, R., Gursky, H., Schwarz, J., and van Paradijs, J.: 1978, *Astrophys. J.* **225**, L59.

Johnston, M., Bradt, H., Doxsey, R., Griffiths, R., Schwartz, D., and Schwarz, J.: 1979, *Astrophys. J.* **230**, L11.

Johnston, M., Bradt, H., and Doxsey, R.: 1979a, preprint.

Johnston, M., Bradt, H., Doxsey, R., Gursky, H., Schwartz, D., and Schwarz, J.: 1978a, preprint.

Jones, C.: 1976, preprint.

Jones, C. and Forman, W.: 1976, *Astrophys. J.* **209**, L131.

Jones, A., Pounds, K., Ricketts, M., Willmore, A., and Morrison, L.: 1972, *Nature* **235**, 152.

Jones, C., Giacconi, R., Forman, W., and Tananbaum, H.: 1974, *Astrophys. J.* **191**, L71.

Jones, C., Forman W., Tananbaum, H., and Turner, M.: 1976, *Astrophys. J.* **210**, L9.

Joss, P.: 1975, *IAU Circ.* No. 2863.

Joss, P. and Rappaport, S.: 1977, *Nature* **265**, 222.

Joss, P., Avni, Y., and Rappaport, S.: 1978, *Astrophys. J.* **214**, 874.

Joss, P., Li, F., Nelson, J., and Middleditch, J.: 1979, preprint.

Julien, P. and Helmken, H.: 1978, *Nature* **272**, 699.

Kaluzienski, L.: 1976, *IAU Circ.* No. 2935.

Kaluzienski, L.: 1977, Thesis.

Kaluzienski, L.: 1977a, *IAU Circ.* No. 3106.

Kaluzienski, L. and Holt, S.: 1977, *IAU Circ.* No. 3144.

Kaluzienski, L. and Holt, S.: 1977a, *IAU Circ.* No. 3099.

Kaluzienski, L. and Holt, S.: 1978, *IAU Circ.* No. 3259.

Kaluzienski, L. and Holt, S.: 1978a, *IAU Circ.* No. 3320.

Kaluzienski, L. and Holt, S.: 1978b, *IAU Circ.* No. 3349.

Kaluzienski, L. and Holt, S.: 1978c, *IAU Circ.* No. 3328.

Kaluzienski, L. and Holt, S.: 1978d, *IAU Circ.* No. 3129.

Kaluzienski, L. and Holt, S.: 1978e, *IAU Circ.* No. 3372.

Kaluzienski, L., Holt, S., Boldt, E., and Serlemitsos, P.: 1975a, *Nature* **256**, 633.

Kaluzienski, L., Holt, S., Boldt, E., Serlemitsos, P., Hadie, G., Pounds, K., Ricketts, M., and Watson, M.: 1975b, *Astrophys. J.* **201**, L121.

Kaluzienski, L., Holt, S., Boldt, E., and Serlemitsos, P.: 1976a, *IAU Circ.* No. 2918.

Kaluzienski, L., Holt, S., Boldt, E., and Serlemitsos, P.: 1967b, *IAU Circ.* No. 2935.

Kaluzienski, L., Holt, S., Boldt, E., and Serlemitsos, P.: 1976c, *Astrophys. J.* **208**, L71.

Kaluzienski, L., Robinson-Saba, J., Boldt, E., Holt, S., Swank, J., Serlemitsos, P., and Rothschild, R.: 1978, *IAU Circ.* No. 3174.

Kasturirangan, K., Rao, U., Sharma, D., and Radha, M.: 1976, *Nature* **260**, 226.

Kelley, R. and Bradt, H.: 1978, *IAU Circ.* No. 3165.

Kellogg, E., Gursky, H., Murray, S., Tananbaum, H., and Giacconi, R.: 1971, *Astrophys. J.* **169**, L99.

Kemp, J.: 1977, *IAU Circ.* No. 3060.

Kemp, J., Herman, L., Rudy, R., and Barbour, M.: 1977, *Nature* **270**, 227.

Kemp, J., Herman, L., and Barbour, M.: 1978, *Astron. J.* **83**, 962.

Kemp, J., Barbour, M., Herman, L., Rand Rudy, R.: 1978a, *Astrophys. J.* **220**, L123.

Kestenbaum, H., Kohen, G., Long, K., Novick, R., Silver, E., Weisskopf, M., and Wolff, R.: 1976, *Astrophys. J.* **208**, L27.

Kestenbaum, H., Ku, W., Long, K., Silver, E., and Novick, R.: 1978, *Astrophys. J.* **262**, 282.

King, A., Ricketts, M., and Warwick, R.: 1979, *Monthly Notices Roy. Astron. Soc.* **187**, 77P.

Kirshner, R.: 1975, *IAU Circ.* No. 2840.

Kleinmann, D.: 1976, *IAU Circ.* No. 2959.

Kleinmann, D., Kleinmann, S., and Wright, E.: 1976a, *Astrophys. J.* **210**, L83.

Kleinmann, S., Joice, R., and Capps, R.: 1976b, in *X-ray Binaries*, p. 355.

Kleinmann, S., Brecher, K., and Ingham, W.: 1976c, *Astrophys. J.* **207**, 532.

Kondo, Y., Parsons, Y., Parsons, S., Henize, K., Wray, J., and Benedict, G.: 1976, in *X-ray Binaries*, p. 551.

Krzeminski, W.: 1974, *Astrophys. J.* **192**, L135.

Kudriavtsev, M.I., Melioranski, A.S., Savenko, I.A., and Yushkov, B.Yu.: 1976, *Astron. Circ.* (Russian) No. 924.

Kurt, V. G., Moskalenko, E. I., Titarchuk, A. G., and Scheffer, E. E.: 1976, *Pis'ma v Astron. J.* (Russian) **2**, 107.

Lamb, R., Dower, R., and Tickle, R.: 1979, *Astrophys. J.* **229**, L19.

Lamb, R. and Worral, D.: 1979, *Astrophys. J.* **231**, L121.

Lamb, F., Pines, D., and Shaham, J.: 1976, in *X-ray Binaries*, p. 141.

Lampton, M., Tuohy, I., Garmire, G., and Charles, P.: 1979, preprint.

Lategan, H.: 1978, *Astron. Astrophys.* **64**, L5.

Lawrence, A., Pye, J., and Elvis, M.: 1977, *Monthly Notices Roy. Astron. Soc.* **181**, 93.

Lea, S., Mason, K., Reichert, G., Charles, P., and Riegler, G.: 1979, *Astrophys. J.* **227**, L67.

Lewin, W.: 1976a, *IAU Circ.* Nos. 2914, 2922.

Lewin, W.: 1976b, *IAU Circ.* Nos. 2911, 2918, 2983.

Lewin, W.: 1977, *Monthly Notices Roy. Astron. Soc.* **179**, 43.

Lewin, W. and Joss, P.: 1977, *Nature* **270**, 211.

Lewin, W., Clark, G., and Smith, W.: 1968a, *Astrophys. J.* **152**, L49.

Lewin, W., Clark, G., and Smith, W.: 1968b, *Canad. J. Phys.* **46**, part 3, S 409.

Lewin, W., McClintock, J., Ryckman, S., and Smith, W.: 1971, preprint.

Lewin, W., Doty, J., and Hoffman, J.: 1976a, *IAU Circ.* No. 2984.

Lewin, W., Hoffman, J., and Doty, J.: 1976b, *IAU Circ.* No. 2994.

Lewin, W., Doty, J., Clark, G., Rappaport, S., Bradt, H., Doxsey, R., Hearn, D., Hoffman, J., Jernigan, J., Li, F., Mayer, W., McClintock, J., Primini, F., and Richardson, J.: 1976c, *Astrophys. J.* **207**, L95.

Lewin, W., Hoffman, J., Doty, J., Hearn, D., Clark, G., Jernigan, J., Li, F., McClintock, J., and Richardson, J.: 1976d, *Monthly Notices Roy. Astron. Soc.* **177**, 83P.

Lewin, W., Li, F., Hoffman, J., Doty, J., Buff, J., Clark, G., and Rappaport, S.: 1976e, *Monthly Notices Roy. Astron. Soc.* **177**, 93P.

Lewin, W., Hoffman, J., Doty, J., Li, F., and McClintock, J.: 1977a, *IAU Circ.* No. 3075.

Lewin, W., Hoffman, J., and Doty, J.: 1977b, *IAU Circ.* No. 3087.

Li, F.: 1976, *IAU Circ.* No. 2936.

Li, F. and Lewin, W.: 1976, *IAU Circ.* No. 2983.

Li, F., Sprott, G., and Clark, G.: 1976, *Astrophys. J.* **203**, 187.

Li, F., Lewin, W., Clark, G., Doty, J., Hoffman, J., and Rappaport, S.: 1977, *Monthly Notices Roy. Astron. Soc.* **179**, 21P.

Li, F., Jernigan, J., and Clark, G.: 1977a, *IAU Circ.* No. 3125.

Li, F., van Paradijs, J., Clark, G., Jernigan, J., Laustsen, S., and Zuiderwijk, E.: 1978, *Nature* **276**, 799.

Li, F., Clark, G., and Rappaport, S.: 1978a, *IAU Circ.* No. 3238.

Liller, W.: 1975, *IAU Circ.* No. 2888.

Liller, W.: 1976a, in *X-ray Binaries*, p. 155.

Liller, W.: 1976b, in *X-ray Binaries*, p. 513.

Liller, W.: 1976c, *IAU Circ*. Nos. 2929, 2936.

Liller, W.: 1978, *IAU Circ*. No. 3176.

Liller, M. and Liller, W.: 1976, *Astrophys. J.* **207**, L109.

Lloyd, C., Noble, R., and Penston, M.: 1977: *Monthly Notices Roy. Astron. Soc.* **179**, 675.

Long, K. and Kestenbaum, H.: 1978, *Astrophys. J.* **226**, 276.

Lozinskaya, T. A.: 1975, *Pis'ma v Astron. J.* (Russian), **1**, 25.

Lucke, R., Yentis, D., Friedman, H., Fritz, G., and Shulman, S.: 1975, *IAU Circ.*, No. 2878.

Lugger, P.: 1978, *Astrophys. J.* **225**, L21.

Lutyi, V. M.: 1976, *Pis'ma v Astron.* J. (Russian) **2**, 402.

Malina, R., Lampton, M., and Bowyer, S.: 1976, *Astrophys. J.* **207**, 894.

Manzo, G., Molteni, D., Robba, N.: 1978, *Astron. Astrophys.* **70**, 317.

Maraschi, L., Huckle, H., Ives, J., and Sanford, P.: 1976, *Nature* **263**, 34.

Maraschi, L., Markert, T., Helmken, H., Apparao, K., Bradt, H., Wheaton, W., and Peterson, L.: 1977, preprint.

Margon, B.: 1976, in *X-ray Binaries*, p. 719.

Margon, B.: 1978, *Astrophys. J.* **219**, 613.

Margon, B.: 1978a, *IAU Circ.* No. 3345.

Margon, B. and Bradt, H.: 1977, *IAU Circ.* No. 3144.

Margon, B., Lampton, M., Bowyer, S., and Cruddace, R.: 1971, *Astrophys. J.* **169**, L23, L45.

Margon, B., Malina, R., Bowyer, S., Cruddace, R., and Lampton, M.: 1976, *Astrophys. J.* **203**, L25.

Markert, T. and Clark, G.: 1975, *Astrophys. J.* **196**, L55.

Markert, T., Canizares, C., Clark, G., Lewin, W., Schnopper, H., and Sprott, G.: 1973, *Astrophys. J.* **184**, L67.

Markert, T., Bradt, H., Clark, G., Lewin, W., Li, F., Schnopper, H., Sprott, G., and Wargo, G.: 1975, *IAU Circ.* No. 2765.

Markert, T., Canizares, C., Clark, G., Li, F., Northridge, P., Sprott, G., and Wargo, G.: 1976a, *Astrophys. J.* **206**, 265.

Markert, T., Backman, D., and McClintock, J.: 1976b, *Astrophys. J.* **208**, L115.

Markert, T., Canizares, C., Clark, G., Hearn, D., Li, F., Sprott, G., and Winkler, P.: 1977, *Astrophys. J.* **218**, 801.

Markert, T., Winkler, P., Laird, F., Clark, G., Hearn, D., Sprott, G., Li, F., Bradt, H., Lewin, W., and Schnopper, H.: 1978, preprint.

Marshall, F., Mushotsky, R., Boldt, E., Holt, S., Rothschild, R., and Serlemitsos P.: 1978, *Nature* **275**, 624.

Marshall, H.: 1978, *IAU Circ.* No. 3336.

Marshall, F. and Jernigan, J.: 1978, *IAU Circ.* No. 3224.

Marshall, F., Boldt, E., Holt, S., Mushotsky, R., Pravdo, S., Rothschild, and R. Serlemitsos, P.: 1979, *Astrophys. J. Suppl.* **40**, 657.

Marshall, H., Ulmer, M., Hoffman, J., Doty, J., and Lewin, W.: 1973, *Astrophys. J.* **227**, 555.

Marshall, H., Ulmer, M., Hoffman, J., Doty, J., and Lewin, W.: 1978a, preprint.

Marshall, F., Mushotzky, R., Boldt, E., Holt, S., and Serlemitsos, P.: 1978b, *IAU Circ.* No. 3314.

Marshall, N. and Watson, M.: 1978, *IAU Circ.* No. 3318.

Martinov, D. Ya.: 1976, *IAU Circ.* No. 2953.

Mason, K. and Culhane, J.: 1978, *Monthly Notices Roy. Astron. Soc.* **185**, 673.

Mason, K., Sanford, P., and Ives, J.: 1976a, in *X-ray Binaries*, p. 255.

Mason, K., Branduardi, G., and Sanford, P.: 1976b, in *X-ray Binaries*, p. 559.

Mason, K., Charles, P., White, N., Culhane, J., Sanford, P., and Strong, K.: 1976c, **177**, 513.

Mason, K., Murdin, P., Parkes, G., and Visvanatham, N., *Monthly Notices Roy. Astron. Soc.* **184**, 45.

Mason, K., Bell, S., and White, N.: 1976d, *IAU Circ.* No. 2932.

Mason, K., Lampton, M., Charles, P., and Bowyer, S.: 1978a, *Astrophys. J.* **226**, L129.

Mason, K., Cordova, F., and Swank, J.: 1978b, preprint.

Matilsky, T.: 1976, *IAU Circ.* No. 2949.

Matilsky, T.: 1977, *Astrophys. J.* **217**, L83.

Matilsky, T. and Jessen, J.: 1978, *IAU Circ.* No. 3193.

Matilsky, T., Giacconi, R., Gursky, H., Kellogg, E., and Tananbaum, H.: 1972, *Astrophys. J.* **174**, L53.

Matilsky, T., Bradt, H., Buff, J., Clarc, G., Jernigan, J., Goss, P., Laufer, B., McClintock, J., and Zubrod, D.: 1976, *Astrophys. J.* **210**, L127.

Matilsky, T., La Sala, J., and Jessen, J.: 1978, *Astrophys. J.* **224**, L119.

Matteson, J., Mushotzky, R., Paciesas, W., and Laros, J.: 1976, in *X-ray Binaries*, p. 407.

Mauder, H.: 1976a, in *X-ray Binaries*, p. 173, 177.

Mauder, H.: 1976b, *IAU Circ.* No. 2946.

Mayer, W.: 1976, *IAU Circ.* No. 3006.

Mazets, E., Golenetskij, S., Il'inskij, V., Panov, V., Aptekar', R., Gur'gan, Yu., Sokolov, I., Sokolova, Z., and Kharitonova, T.: 1979, *Pis'ma v Astron. J.* **5**, 307.

McClintock, J.: 1977, *IAU Circ.* No. 3084.

McClintock, J. and Canizares, C.: 1978, *IAU Circ.* No. 3250.

McClintock, J., van Paradijs, J., Remillard, R., and Canizares, C., Koski, A.: 1979, preprint.

McClintock, J. and Rappaport, S.: 1976, in *X-ray Binaries*, p. 661.

McClintock, J., Rappaport, S., Joss, P., Bradt, H., Buff, J., Clark, G., Hearn, D., Lewin, W., Matilsky, T., Mayer, W., and Primini, F.: 1976, *Astrophys. J.* **206**, L99.

McClintock, J., Li, F., Nugent, J., and Rappaport, S.: 1977a, *IAU Circ.* No. 3039.

McClintock, J., Canizares, C., Bradt, H., Doxsey, R., Jernigan, J., and Hiltner, W.: 1977b, preprint.

McClintock, J., Canizares, C., and Backman, D.: 1978, preprint.

McClintock, J., Grindlay, J., Canizares, C., van Paradijs, J., Cominsky, L., Li, F. and Lewin, W.: 1979a, preprint.

McConnell, D. and Cowley, A.: 1972, *IAU Circ.* No. 2401.

McHardy, I.: 1978, *Monthly Notices Roy. Astron. Soc.* **184**, 783.

Meekins, J., Henry, R., Fritz, G., Friedman, H., and Byram, E.: 1969, *Astrophys. J.* **157**, 197.

Melnick, J. and Quintana, H.: 1975, *Astrophys. J.* **198**, L97.

Mewe, R., Heise, J., Gronenschild, E., Brinkmann, A., Schrijver, J., and den Boggende, A.: 1975, *Nature* **256**, 712.

Meyer, W., Bradt, H., and Rappaport, S.: 1970, *Astrophys. J.* **159**, L115.

Middleditch, J. and Nelson, J.: 1975, preprint.

Milgrom, M.: 1976, *Astron. Astrophys.* **51**, 215.

Milgrom, M.: 1978, *Astron. Astrophys.* **65**, L1.

Miller, H.: 1979, *Astrophys. J.* **227**, 52.

Mironov, A. V. and Tscherepaschuk, A. M.: 1976, *Astron. Circ.* (Russian) No. 913.

Moore, W. and Garmire, G.: 1976, *Astrophys. J.* **206**, 247.

Moskalenko, B. I., Kurt, V. G., Scheffer, E. K., Titarchuk, L. C., and Golovanov, I. A.: 1976, *Pis'ma v Astron. J.* (Russian) **2**, 528.

Murdin, P., Penston, M., Penston, M., Class, I., Sanford, P., Hawkins, F., Mason, K., and Willmore, A.: 1974, *Monthly Notices Roy. Astron. Soc.* **169**, 25.

Murdin, P., Penston, M., and Penny, A.: 1976, *Monthly Notices Roy. Astron. Soc.* **176**, 233.

Murdin, P., Griffiths, R., Pounds, K., Watson, M., and Longmore, A.: 1977, *Monthly Notices Roy. Astron. Soc.* **178**, 27P.

Murdin, P., Morton, O., and Thomas, R.: 1979, *Monthly Notices Roy. Astron. Soc.* **186**, 43P.

Murray, S. and Ulmer, M.: 1976, *Astrophys. J.* **210**, 230.

Murray, S.: 1977, preprint.

Mushotzky, R., Serlemitsos, P., Smith, R., Boldt, E., and Holt, S.: 1978, *Astrophys. J.* **225**, 21.

Naranan, S., Shulman, S., Friedman, H., and Fritz, G.: 1976, *Astrophys. J.* **208**, 718.

Naranan, S., Shulman, S., Yentis, D., Fritz, G., and Friedman, H.: 1977, *Astrophys. J.* **213**, L53.

Natali, G. and Messi, R.: 1978, *Astron. Astrophys.* **67**, L33.

Newell, B., da Costa, G., and Norris, J.: 1976, *Astrophys. J.* **208**, L55.

Nishimura, J., Fujii, M., Tawara, Y., Oda, M., Ogawara, Y., Yamagami, T., Miyamoto, S., Kajiwara, M., Murakami, H., Yoshimori, M., Nakagawa, M., and Sakurai, T.: 1978, *Nature* **272**, 337.

Novikova, G. V., Erochina, E. V., Kurt, V. G., Moskalenko, E. I., Titarchuk, L. G., and Scheffer, E. K.: 1977, *Kosmich. Issledov.* (Russian) **15**, 321.

Nugert, J. and Garmire, G.: 1978, *Astrophys. J.* **226**, L83.

Oda, M.: 1978, *IAU Circ.* No. 3349.

Ogelman, H., Beuormann, K., Kanbach, G., Mayer-Hasselwander, H., Capozzi, D., Fiordilino, E., and Molteni, D.: 1977, *Astron. Astrophys.* **58**, 385.

Oke, J.: 1976, *Astrophys. J.* **209**, 547.

Olson, E.: 1977, *Astrophys. J.* **215**, 166.

Osmer, P., Hiltner, W., and Whelan, J.: 1975, *Astrophys. J.* **195**, 705.

Owen, F., Balonel, T., Dickey, J., Terzian, Y., and Gottesman, S.: 1976, *Astrophys. J.* **203**, L15.

Pacheco, J.: 1978, *Astron. Astrophys.* **70**, L49.

Pakull, M.: 1978, *IAU Circ.* No. 3317.

Paradijs, van J.: 1975, *IAU Circ.* No. 2841.

Paradijs, van J., Hammerschlag-Hensberge, G., van den Heuvel, E., Takens, R., Zuiderwijk, E., and de Loore, G.: 1976, in *X-ray Binaries*, p. 643.

Pakull, M.: 1978, *ESO Messenger*, Nos. 13, 17.

Palmieri, T., Burginyon, G., Hill, R., Scudder, J., and Seward, F.: 1972, *Astrophys. J.* **177**, 387.

Parenago, P. P.: 1954, *Stellar Astronomy* in (Russian).

Parkes, G., Charles, P., Culhame, L., and Ives, J.: 1977, *Monthly Notices Roy. Astron. Soc.* **179**, 55.

Parkes, G., Murdin, P., and Mason, K.: 1978, *IAU Circ.* No. 3184.

Parkes, G., Murdin, P., and Mason, K.: 1978a, *Monthly Notices Roy. Astron. Soc.* **184**, 73P.

Parsignault, D., Grindlay, J., Schnopper, H., Scherer, E., and Gursky, H.: 1976a, preprint.

Parsignault, D., Scheier, E., Grindlay, J., Schnopper, H., and Gursky, H.: 1976b, in *X-ray Binaries*, p. 267.

Parsignault, D., Schreier, E., Grindlay, J., and Gursky, H.: 1976c, *Astrophys. J.* **209**, L73.

Parsignault, D. and Grindlay, J.: 1978, *Astrophys. J.* **225**, 970.

Patterson, J., Robinson, E., and Kiplinger, A.: 1978, *Astrophys. J.* **226**, L137.

Pedersen, A.: 1976, *IAU Circ.* No. 2972.

Penny, A., Penford, J., and Balona, L.: 1975, *Monthly Notices Roy. Astron. Soc.* **171**, 387.

Penston, M., Murdin, P., Penston, M., Bingham, R., and Sinclair, J.: 1976, *Monthly Notices Roy. Astron. Soc.* **176**, 237.

Perry, M. and Peterson, B.: 1974, *Astron. J.* **79**, 1.

Persi, P., Ferrary-Toniolo, M., Spada, G., Conti, G., Di Benedetto, P., Tanzi, E., and Tarenghi, M.: 1979, *Monthly Notices Roy. Astron. Soc.* **187**, 293.

Perch, P.: 1976, *IAU Circ.* No. 3023.

Peterson, B.: 1975, *IAU Circ.* No. 2837.

Petro, L.: 1978, *IAU Circ.* No. 3327.

Pietsch, W., Kendiziorra, E., Staubert, R., and Trümper, J.: 1976, in *X-ray Binaries*, p. 277.

Pineda, F. and Schnopper, H.: 1978, *IAU Circ.* No. 3190.

Pizzichini, G., Palumbo, G., and Spizzichino, A.: 1975, *Astrophys. J.* **195**, L1.

Polidan, R., Sanford, P., White, N., Pollard, G., and Locke, M.: 1978, *IAU Circ.* No. 3234.

Polidan, R., Pollard, G., Locke, M., and Sanford, P.: 1978a, *Nature* **275**, 296.

Pravdo, S., Becker, R., Boldt, E., Holt, S., Rothschild, R., Serlemitsos, P., and Swank, J.: 1976, *Astrophys. J.* **208**, L67.

Pravdo, S., Becker, R., Boldt, E., Holt, S., Serlemitsos, P., and Swank, J.: 1977, *Astrophys. J.* **215**, L61.

Pravdo, S., Boldt, E., Holt, E., Rothschild, R., Serlemitsos, P.: 1978, *Astrophys. J.* **225**, 53.

Pravdo, S., Bussard, R., Becker, R., Boldt, E., Holt, S., and Serlemitsos, P.: 1978a, *Astrophys. J.* **225**, 988.

Price, R., Seward, F., and Swift, C.: 1972, *Astrophys. J.* **176**, 611.

Priedhorsky, W.: 1977, *Astrophys. J.* **212**, L117.

Priedhorsky, W. and Krzeminsky, W.: 1978, *Astrophys. J.* **219**, 597.

Primini, F., Rappaport, S., Joss, P., Clark, G., Lewin, W., Li, F., Mayer, W., and McClintock, J.: 1976, *Astrophys. J.* **210**, L71.

Primini, F., Cooke, B., Dolson, C., Howe, S., Scheepmaker, A., Wheaton, W., Lewin, W., Baity, W., Gruber, D., Matteson, J., and Peterson, L.: 1976, preprint.

Proctor, R., Skinner, G., and Willmore, A.: 1978, *Monthly Notices Roy. Astron. Soc.* **185**, 745.

Protheroe, R. and Wolfendale, A.: 1980, *Astron. Astrophys.* **84**, 128.

Pye, J. and Cooke, B.: 1976, *Nature* **260**, 410.

Pye, J., Elvis, M., and Eadie, J.: 1976, *IAU Circ.* No. 2985.

Rao, U., Chitnis, E., Prakasarao, A., and Jayanthi, U.: 1969, *Astrophys. J.* **157**, L127.

Rappaport, S.: 1975, *IAU Circ.* No. 2869.

Rappaport, S., Bradt, H., Naranan, S., and Spada, A.: 1969, *Nature* **221**, 428.

Rappaport, S., Doxsey, R., Solinger, A., and Borken, R.: 1974a, *Astrophys. J.* **194**, 329.

Rappaport, S., Cash, W., Doxsey, R., McClintock, J., and Moore, G.: 1974b, *Astrophys. J.* **187**, L5.

Rappaport, S. and McClintock, J.: 1975, *IAU Circ.* No. 2833.

Rappaport, S., Levine, A., Doxsey, R., and Bradt, R.: 1975, *Astrophys. J.* **196**, L15.

Rappaport, S., Joss, P., Bradt, H., Clark, G., and Jernigan, J.: 1976a, *Astrophys. J.* **208**, L119.

Rappaport, S., Buff, J., Clark, G., Lewin, W., Matilsky, W., and McClintock, J.: 1976b, *Astrophys. J.* **206**, L139.

Rappaport, S., Clark, G., Dower, R., Doxsey, R., Jernigan, J., and Li, F.: 1977a, *Nature* **268**, 705.

Rappaport, S., Markert, T., Li, F., Clark, G., Jernigan, J., and McClintock, J.: 1977b, preprint.

Rappaport, S., Clark, G., Cominsky, L., Joss, P., and Li, F.: 1978a, *Astrophys. J.* **224**, L1.

Rappaport, S., Clark, G., Cominsky, L., and Li, F.: 1978b, *IAU Circ.* No. 3171.

Reina, C.: 1974, Mem. Soc. Astron. Ital., **44**, 603.

Ricker, G.: 1977, preprint.

Ricketts, M.: 1978, *Monthly Notices Roy. Astron. Soc.* **183**, 51.

Ricketts, M. and Cooke, B.: 1977, *IAU Circ.* No. 3039.

Ricker, G., Schoenmaker, A., Ballintine, J., Doty, J., Kriss, G., Byckman, S., and Lewin, W.: 1975, preprint.

Ricker, G., Gerassimenko, M., McClintock, J., Ryckmann, S., and Lewin, W.: 1976a, *Astrophys. J.* **207**, 333.

Ricker, G., Scheepmaker, A., Ballintine, J., Doty, J., Kriss, G., Byckmann, S., and Lewin, W.: 1977b, in *X-ray Binaries*, p. 279.

Ricker, G. and Primini, F.: 1977, *IAÚ Circ.* No. 3078.

Ricker, G., Clark, G., Doxsey, R., Dower, R., Jernigan, J., Delvaille, J., McAlpine, G., and Hjellming, R.: 1977, preprint.

Ricker, G., Clark, G., Doxsey, R., Dower, R., and Jernigan, J.: 1977a, preprint.

Ricketts, M., Turner, M., Page, C., and Pounds, K.: 1975a, *Nature* **265**, 631.

Ricketts, M., Pounds, K., and Turner, M.: 1975b, *Nature* **257**, 657.

Ricketts, M., Gowan, R., and Warwick, R.: 1978, *IAU Circ.* No. 3238.

Riegler, G., Agrawal, P., and Rosker, M.: 1978, *IAU Circ.* No. 3261.

Riegler, G., Agrawal, P., and Mushotzky, R.: 1978a, preprint.

Robinson, E. and Affricano, J.: 1975, *IAU Circ.* No. 2869.

Rossiger, S.: 1976, *Mitt. Veränderl. Sterne* **7**, 105.

Rotherflug, R., Rocchia, R., and Casse, M.: 1979, *Astrophys. J.* **229**, 669.

Rothschild, R., Boldt, E., Holt, S., and Serlemitsos, P.: 1976, in *X-ray Binaries*, p. 443.

Rowan-Robinson, M. and Fabian, A.: 1975, *Monthly Notices Roy. Astron. Soc.* **170**, 199.

Ryter, C.: 1976, in *X-ray Binaries*, p. 189.

Saden, D., Meidav, M., Evans, W., Byram, E., Chubb, T., and Friedman, H.: 1979, *Nature* **278**, 436.

Sagdeev, R. Z.: 1976, *IAU Circ.* No. 2959.

Samini, J., Share, G., Wood, K., Yentis, D., Meekins, J., Evans, W., Shulman, S., Byram, E., Chubb, T., Friedman, H.: 1979, *Nature* **278**, 434.

Sanduleak, N. and Dolan, J.: 1974, *Astrophys. J.* **187**, L73.

Scharlemann, E.: 1978, *Astrophys. J.* **219**, 617.

Schmidt, G. and Romanishin, W.: 1975, *IAU Circ.* No. 2873.

Schlosser, W. and van Paradijs, J.: 1979, *Astron. Astrophys.* **75**, 112.

Schnopper, H. and Delvaille, J.: 1977, preprint.

Schnopper, H., Delvaille, J., Epstein, A., Helmken, H., Murray, S., Clark, G., Jernigan, J., and Doxsey, R.: 1976, *Astrophys. J.* **210**, L75.

Schnopper, H., Epstein, A., Delvaille, J., Tucker, W., Doxsey, R., and Jernigan, J.: 1977, *Astrophys. J.* **215**, L7.

Schreier, E.: 1977, preprint.

Schreier, E. and Fabbiano, G.: 1976, in *X-ray Binaries*, p. 197.

Schwartz, D.: 1978, *Astrophys. J.* **220**, 8.

Schwartz, D.: 1979, preprint.

Schwartz, D., Bradt, H., Briel, U., Dower, R., Doxsey, R., Fabbiano, G., Griffiths, R., Gursky, H., Johnston, M., Leach, R., Liller, W., Ramsey, A., and Schwarz, J.: 1978, preprint.

Schwartz, D., Gursky, H., Schwartz, J., Bradt, H., and Doxsey, R.: 1978a, preprint.

Schwartz, D., Doxsey, R., Griffiths, R., Johnston, M., and Schwarz, J.: 1978b, preprint.

Schwartz, D., Bradt, H., Doxsey, R., Griffiths, R., Gursky, H., Johnston, M., and Schwarz, J.: 1978c, preprint.

Schwartz, D., Bleach, R., Boldt, E., Holt, S., and Serlemitsos, P.: 1972, *Astrophys. J.* **173**, L51.

Schwartz, D., Bradt, H., Briel, U., Doxsey, R., Fabbiano, G., Griffiths, R., Johnston, M., and Margon, B.: 1979, preprint.

Schwarz, J., Briel, U., Doxsey, R., Fabbiano, G., Griffiths, R., Johnston, M., Schwartz, D., and McKee, J.: 1979, preprint.

Schwarz, J., Briel, U., Doxsey, R., Fabbiano, G., Griffiths, R., Johnston, M., Schwartz, D., McKee, J., and Mushotzky, K.: 1979a, *Astrophys. J.* **213**, L105.

Searle, L.: 1975, *IAU Circ.* No. 2840.

Seaquist, E., Garrison, R., Gregory, P., Taylor, A., Crane, P.: 1979, *Astron. J.* **84**, 1037.

Seward, F.: 1970, preprint.

Seward, F., Burginyon, G., Grader, R., Hill, R., and Palmieri, T.: 1972, *Astrophys. J.* **178**, 131.

Seward, F., Page, C., Turner, M., and Pounds, K.: 1976a, *Monthly Notices Roy. Astron. Soc.* **177**, p. 13.

Seward, F., Page, C., Turner, M., and Pounds, K.: 1976b, *Monthly Notices Roy. Astron. Soc.* **175**, 39P.

Shakura, N. I.: 1975, *Pis'ma v Astron. J.* (Russian), **1**, Nos. 11, 23.

Sims, M. and Watson, M.: 1978, *IAU Circ.* No. 3227.

Share, G., Wood, K., Meekins, J., and Shulman, S.: 1978, *IAU Circ.* No. 3169.

Share, G., Wood, K., Byram, E., Meekins, J., Shulman, S., and Griffiths, R.: 1978a, *IAU Circ.* No. 3197.

Shulka, P. and Wilson, B.: 1970, *Nature* **228**, 1077.

Shulka, P. and Wilson, B.: 1971, *Nature Phys. Sci.* **234**, 149.

Shulman, S., Friedman, H., Henry, R., and Davidson, A.: 1976, in *X-ray Binaries*, p. 127.

Smith, A.: 1976, *IAU Circ.* No. 2980.

Smith, A.: 1978, *Monthly Notices Roy. Astron. Soc.* **182**, 39.

Snyder, W., Davidsen, A., Henry, R., Shulman, S., Fritz, G., and Friedman, H.: 1978, *Astrophys. J.* **222**, L13.

Soderblom, D.: 1976, *IAU Circ.* No. 2971.

Solomon, P. and Sanders, D.: 1978, preprint.

Steiner, J.: 1977, *IAU Circ.* No. 3107.

Stier, M. and Liller, W.: 1976, *Astrophys. J.* **206**, 257.

Staubert, R., Kendziorra, E., Pietsch, W., Reppin, C., Trumper, J., and Voges, W.: 1978, *Astrophys. J.* **225**, 113.

Surdej, J.: 1978, *IAU Circ.* No. 3290.

Swanenburg, B., Bennet, K., Bignami, G., Caraveo, P., Hermson, W., Kanbach, G., Manson, J., Mayer-Hasselwander, H., Paul, J., Sacco, B., Scarsi, B., and Wills, R.: 1978, *Nature* **275**, 298.

Swank, J., Becker, R., Pravdo, S., and Serlemitsos, P.: 1976a, *IAU Circ.* No. 2963.

Swank, J., Becker, R., Pravdo, S., Saba, J., and Serlemitsos, P.: 1976b, *IAU Circ.* No. 3000.

Swank, J., Becker, R., Pravdo, S., Saba, J., and Serlemitsos, P.: 1976c, *IAU Circ.* No. 3010.

Swank, J., Becker, R., Boldt, E., Holt, S., Pravdo, S., Rothschild, R., and Serlemitsos, P.: 1976d, in *X-ray Binaries*, p. 207.

Swank, J., Becker, R., Boldt, E., Holt, S., Pravdo, S., Rothschild, R., and Serlemitsos, P.: 1976e, *Astrophys. J.* **209**, L57.

Swank, J., Becker, R., Boldt, E., Holt, S., Pravdo, S., and Serlemitsos, P.: 1977, *Astrophys. J.* **212**, L73.

Swank, J., Becker, R., Boldt, E., Holt, S., and Serlemitsos, P.: 1978a, *Monthly Notices Roy. Astron. Soc.* **182**, 349.

Swank, J., Boldt, E., Holt, S., Rothschild, R., and Serlemitsos, P.: 1978b, *Astrophys. J.* **226**, L133.

Szkody, P. and Brownlee, D.: 1977, *Astrophys. J.* **212**, L113.

Takagishi, K., Nagareda, K., Matsuoka, M., Fujii, M., Sato, S., van Paradijs, J., Hoffman, J., Jernigan, G., Wheaton, W., and Lewin, W.: 1978, preprint.

Tananbaum, H., Chaisson, L., Forman, W., and Jenss, C.: 1976, *Astrophys. J.* **209**, L125.

Tapia, S.: 1977, *Astrophys. J.* **212**, L125.

Thomas, R., Davison, P., Claney, M., and Busselli, G.: 1975, *Monthly Notices Roy. Astron. Soc.* **170**, 569.

Thomas, R., Greenhill, J., and Watts, D.: 1978, *IAU Circ.* No. 3334.

Thompson, D., Fichtel, C., Kniffen, D., and Ogelman, H.: 1977, *Astrophys. J.* **214**, L17.

Thorstensen, J., Charles, P., and Bowyer, S.: 1978, *Astrophys. J.* **220**, L131.

Thorstensen, J., Charles, P., and Bowyer, S.: 1978a, *IAU Circ.* No. 3253.

Thorstensen, J., Charles, P., Bowyer, S., Briel, U., Doxsey, R., Griffirths, R., and Schwartz, D.: 1979, preprint.

Toor, A.: 1977, *Astrophys. J.* **215**, L57.

Topka, K., Fabricant, D., Karnden, F., Gorenstein, P., Rosner, R.: 1979, *Astrophys. J.* **229**, 661.

Tramper, J., Pietsch, W., Reppin, C., and Voges, W.: 1978, *Astrophys. J.* **219**, L105.

Tsikoudi, V. and Hudson, H.: 1975, *Astrophys. Astron.* **44**, 273.

Tuohy, I.: 1976, in *X-ray Binaries*, p. 219.

Tuohy, I., Cordova, F., Garmire, G., Nugent, J., Charles, P., Bowyer, S., and Watter, F.: 1978, *IAU Circ.* No. 3247.

Tuohy, I., Mason, K., Clark, D., Cordova, F., Charles, P., Watter, F., and Garmire, G.: 1979, *Astrophys. J.* **230**, L27.

Tuohy, I., Lamb, F., Garmire, G., and Mason, K.: 1978a, preprint.

Ulmer, M. and Jernigan, J.: 1978, preprint.

Ulmer, M., Baily, W., Wheaton, W., and Peterson, L.: 1973, *Astrophys. J.* **184**, L117.

Ulmer, M., Murray, S., and Gursky, H.: 1976, *Astrophys. J.* **208**, 47.

Ulmer, M., Lewin, W., Hoffman, J., Doty, J., and Marshall, H.: 1977, *Astrophys. J.* **214**, L11.

Ulmer, M., Hjellming, R., Lewin, W., Hoffman, J., Jernigan, J., Wheaton, W., Primini, F., Doty, J., and Marshall, H.: 1978, *Nature* **276**, 799.

van den Bergh, S.: 1976, *Astrophys. J.* **208**, L17.

van Genderen, A.: 1979, *Astron. Astrophys.* **73**, 183.

van Genderen, A.: 1977, *Astron. and Astrophys.* **54**, 733.

van Paradijs, J., Joss, P., Cominsky, L., and Lewin, W.: 1979, preprint.

van Paradijs, J., Cominsky, L., and Lewin, W.: 1978, *IAU Circ*. No. 3294.

van Paradijs, J., Cominsky, L., and Lewin, W.: 1978a, preprint.

van Speybroeck, L., Epstein, A., Forman, W., Giacconi, R., Jones, C., Liller, W., and Smarr, L.: 1979, *Astrophys. J.* **234**, L45.

Vedrenne, G., Zenchenko, V., Kurt, V., Niel, M., Hurley, K., and Estulin, I.: 1979, *Pis'ma v Astron. J.* **5**, 588.

Vidal, N.: 1973, *Astrophys. J.* **186**, L81.

Vidal, N.: 1976, in *X-ray Binaries*, p. 575.

Villa, G., Page, C., Turner, M., Cooke, B., Ricketts, M., Pounds, K., and Adams, D.: 1976, *Monthly Notices Roy. Astron. Soc.* **176**, 609.

Wade, R. and Oke, J.: 1977, *Astrophys. J.* **215**, 568.

Walker, E.: 1976, in *X-ray Binaries*, p. 569.

Walter, E.: 1976, *IAU Circ*. No. 2959.

Walter, F., Charles, P., and Bowyer, S.: 1978a, *Astron. J.* **83**, 1539.

Walter, F., Charles, P., and Bowyer, S.: 1978b, *Nature* **274**, 569.

Walter, F., Charles, P., and Bowyer, S.: 1978c, *Astrophys. J. J.* **225**, L119.

Walter, F., Charles, P., Bowyer, S., and Garmire, G.: 1978, *IAU Circ*. No. 3173.

Walter, F., Charles, P., and Thornstensen, J.: 1978d, *IAU Circ*. No. 3243.

Ward, W., Allen, D., Smith, M., and Wright, A.: 1979, *IAU Circ*. No. 3335.

Ward, M., Penston, M., Murray, S., and Clements, E.: 1975, *Nature* **257**, 659.

Watson, M.: 1976, *IAU Circ*. No. 2934.

Watson, M. and Ricketts, M.: 1978, *Monthly Notices Roy. Astron. Soc.* **183**, 35P.

Watson, M. and Pye, J.: 1977, *IAU Circ*. No. 3078.

Watson, M., Pye, J., Elvis, M., and Lawrence, A.: 1976, *IAU Circ*. No. 3013.

Webster, B.: 1973, *Monthly Notices Roy. Astron. Soc.* **164**, 381.

Weiskopf, M., Cohen, G., Kestenbaum, H., Long, K., Novick, R., and Wolff, R.: 1976, *Astrophys. J.* **208**, L125.

Wheaton, W., Ulmer, M., Baity, W., Datlowe, D., Elcan, M., Peterson, L., Klebesadel, R., Strong, I., Cline, T., and Desai, U.: 1973, *Astrophys. J.* **185**, L57.

Wheaton, W.: 1975, *IAU Circ*. No. 2761.

White, N.: 1977, *IAU Circ*. No. 3118.

White, N. and Carpenter, G.: 1978, *Monthly Notices Roy. Astron. Soc.* **183**, 11P.

White, N. and Mason, K.: 1977, *Nature* **267**, 229.

White, G. and Ricketts, M.: 1977, *Astrophys. J.* **18**, L79.

White, N., Huckle, H., Mason, K., Charles, P., Pollard, G., Culhane, J., and Sanford, P.: 1975, *IAU Circ*. No. 2870.

White, N., Mason, K., Sanford, P., and Murdin, P.: 1976a, *Monthly Notices Roy. Astron. Soc.* **176**, 201.

White, N., Mason, K., Huckle, H., Charles, P., and Sanford, P.: 1976b, *Astrophys. J.* **209**, L119.

White, N., Mason, K., Sanford, P., Ilovaisky, S., and Chevalier, C.: 1976c, *Monthly Notices Roy. Astron. Soc.* **176**, 91.

White, N., Mason, K., and Sanford, P.: 1976d, *IAU Circ*. No. 2920.

White, N., Mason, K., and Sanford, P.: 1978, *Monthly Notices Roy. Astron. Soc.* **184**, 67P.

White, N., Sanford, P., and Weiler, E.: 1978a, *Nature* **274**, 569.

White, N., Pravdo, S., and Swank, J.: 1978b, *IAU Circ*. No. 3342.

White, N., Mason, K., Sanford, P., Johnson, H., and Catura, R.: 1978c, *Astrophys. J.* **220**, 600.

White, N., Burnell, S., and Carpenter, G.: 1977, *IAU Circ*. No. 3067.

White, N., Davison, J., and Carpenter, G.: 1977a, *IAU Circ*. No. 3099.

Willmore, A., Mason, K., Sanford, P., Hawkins, F., Murdin, P., Penston, M., and Penston, M.: 1974, *Monthly Notices Roy. Astron. Soc.* **169**, 7.

Willmore, A.: 1975, *IAU Circ*. No. 2876.

Willmore, A.: 1977, in *Highlights of Astronomy*, Vol. 4, Part I, 87–94.

Wilson, A. and Carpenter, G.: 1976, *Nature* **261**, 295.

Wilson, A., Ward, M., Axon, D., Elvis, M., and Meurs, E.: 1979, *Monthly Notices Roy. Astron. Soc.* **187**, 109.

Winkler, P.: 1978, preprint.

Winkler, P.: 1978a, preprint.

Winkler, P., Frank, J., and Clark, G.: 1974, *Astrophys. J.* **191**, L67.

Winkler, P. and Laird, F.: 1976, *Astrophys. J.* **204**, L111.

Winkler, P., Hearn, D., Richardson, J., and Behnken, J.: 1978, *CSK-P*. 78–64.

Winkler, P., Hearn, D., Richardson, J., and Behnken, J.: 1979, *Astrophys. J.* **229**, L123.
Woltjer, L.: 1972, *Ann. Rev. Astron. and Astrophys.* **10**, 129.
Woltjer, L. and Mason, K.: 1979, preprint.
Wood, K., Share, G., Shulman, S., and Meekins, J.: 1978, *IAU Circ.* No. 3169.
Wood, K., Share, G., Johnston, N., Yentis, D., and Byram, E.: 1978, *IAU Circ.* No. 3203.
Wyckoff, S. and Wehinger, P.: 1975, *IAU Circ.* No. 2835.
Zuiderwijk, E.: 1978, *IAU Circ.* No. 3221.
Zuiderwijk, E.: 1978a, *IAU Circ.* No. 3250.